YOUNG MATHEMATICIANS
AT WORK

YOUNG MATHEMATICIANS
AT WORK

CONSTRUCTING ALGEBRA

CATHERINE TWOMEY FOSNOT
BILL JACOB

HEINEMANN
Portsmouth, NH

NATIONAL COUNCIL OF
TEACHERS OF MATHEMATICS
Reston, VA

Heinemann
361 Hanover Street
Portsmouth, NH 03801–3912
www.heinemann.com

Offices and agents throughout the world

Published simultaneously by Heinemann and National Council of Teachers of Mathematics.

This material is supported in part by the National Science Foundation under Grant No. DRL-0822034. Any opinions, findings, and conclusions or recommendations expressed in this material are those of the authors and do not necessarily reflect the views of the National Science Foundation.

The authors and publisher wish to thank those who have generously given permission to reprint borrowed material:

Excerpts from *The Box Factory* by Miki Jensen and Catherine Twomey Fosnot, part of the *Contexts for Learning Mathematics Series*. Copyright © 2007 by Catherine Twomey Fosnot. Published by *first*hand, an imprint of Heinemann. Reprinted by permission of Heinemann.

Excerpts from *Beads and Shoes, Making Twos* by Madeline Chang and Catherine Twomey Fosnot, part of the *Contexts for Learning Mathematics Series*. Copyright © 2007 by Catherine Twomey Fosnot. Published by *first*hand, an imprint of Heinemann. Reprinted by permission of Heinemann.

Excerpts from *Trades, Jumps, and Stops* by Catherine Twomey Fosnot and Patricia Lent, part of the *Contexts for Learning Mathematics Series*. Copyright © 2007 by Catherine Twomey Fosnot. Published by *first*hand, an imprint of Heinemann. Reprinted by permission of Heinemann.

Excerpts from *The California Frog-Jumping Contest* by Bill Jacob and Catherine Twomey Fosnot, part of the *Contexts for Learning Mathematics Series*. Copyright © 2007 by Catherine Twomey Fosnot. Published by *first*hand, an imprint of Heinemann. Reprinted by permission of Heinemann.

Portions of Chapter 7 originally appeared in *Connect* magazine, Vol. 19, No. 3 (January/February 2006). Reprinted by permission of Synergy Learning International, Inc.

Figure 8.1: Image from *Comparing Quantities: Britannica Mathematics in Context Work Pages* by Margaret R. Meyer and Margaret A. Pligge. Copyright © 1998 by Encyclopedia Britannica Educational Corporation. Published by Encyclopedia Britannica, Inc. Reprinted by permission of the publisher.

Library of Congress Cataloging-in-Publication Data
Fosnot, Catherine Twomey.
  Young mathematicians at work : constructing algebra / Catherine
Twomey Fosnot, Bill Jacob.
      p.   cm. — (Young mathematicians at work)
  Includes bibliographical references and index.
  ISBN 13: 978-0-325-02841-5
  ISBN 10: 0-325-02841-9
  1. Algebra—Study and teaching (Elementary).   I. Jacob, Bill.   II. Title.
III. Title: Constructing algebra.
  QA159.F67 2010
  372.7—dc22   2009045719

Editor: Victoria Merecki
Production: Elizabeth Valway
Cover design: Bernadette Skok
Composition: Cape Cod Compositors, Inc.
Manufacturing: Valerie Cooper

Printed in the United States of America on acid-free paper

14 13 12 11 10 VP 2 3 4 5

CONTENTS

ACKNOWLEDGMENTS

The two names on the cover of this book mean only that we are the ones who finally sat down at the keyboard. The ideas included here grew out of collaboration between many researchers at the Freudenthal Institute, faculty and staff of Mathematics in the City, and many talented, hard-working teachers who were interested and willing to try things out in their classrooms and examine with us their children's ideas.

First and foremost, we thank our Dutch colleague, Maarten Dolk, who co-authored with Cathy the first three books on number and operation in this series. Many of the ideas written about there have been instrumental in helping us define, describe, and consider our views on the teaching and learning of algebra. He also read portions of this manuscript and provided helpful comments along the way. In particular he contributed to our think tank over a three-year period on algebra as we worked to design sequences of investigations to try in classrooms.

Other members of the think tank were Aad Goddijn, Frans van Galen, Koeno Gravemeijer, Ed Wall, Eve Torrence, Ruben Farley, Loren Pitt, Patricia Lent, Antonia Cameron, Sherrin Hersch, Dawn Selnes, Nina Liu, Miki Jensen, Debbie Katzburg, Despina Stylianou, Lynn Tarlow-Hellman, and Madeline Chang. We gratefully acknowledge their contributions. In particular we wish to thank Patricia Lent, Miki Jensen, and Madeline Chang for field-testing activities, co-authoring the specific units of study described within, and for their willingness to tell their stories, and Debbie Katzburg for reading and commenting on the manuscript along the way. *The California Frog-Jumping Contest* sequence was field-tested in Santa Barbara, California, by Janice Ulloa-Brown. We are extremely grateful for the time she gave to the project, her professionalism, and dedication to the work. Sadly she passed away before this manuscript was complete.

Along the way we have had several funding sources, and we gratefully acknowledge their financial contribution and generosity. In the fall of 2008 The Educational Advancement Foundation in Austin, Texas, funded release time for both of us to work on the manuscript while Bill spent a portion of his sabbatical at Mathematics in the City working with Cathy. NSF has been a consistent funder of our research as well, and recently awarded us funding to film some of the sequences described within (*Trades, Jumps, and Stops* and *The California Frog-Jumping Contest*) and to develop and research professional development materials on the teaching of algebra.

Last, we thank our editors at Heinemann, Victoria Merecki and Alan Huisman, who always read with an insightful eye, offering helpful suggestions and edits to bring clarity and polish to our writing.

PREFACE

This book is the fourth volume in a series of four. The first three books in the Young Mathematicians at Work series were on number and operation and were co-authored with Maarten Dolk. This volume extends that work to encompass the development of algebra in children between the ages of six and twelve.

The series is a culmination of a long and fruitful journey characterized by collaboration, experimentation, reflection, and growth. Mathematics in the City (www.mitcccny.org) was established in 1995 with initial funding from the Exxon-Mobil Foundation and the National Science Foundation. Today, funded by many sources—including continued funding from NSF—it is a large center of in-service and research for mathematics education, K–8. Over the past twenty years, many researchers, mathematicians, teacher educators, and classroom teachers have contributed to our work. Besides the Young Mathematicians at Work series, we have published professional development materials (comprising digital materials and facilitator guides) and a K–6 curriculum (comprised of twenty-four units, colorful posters, and eight trade books). For further information on these, please see www.newperspectivesonlearning.com.

THE IMPORTANCE OF NUMERACY AND ALGEBRA . . .

Numeracy and algebra are critical issues. In today's world children need a strong understanding of number and operation. They need good mental arithmetic strategies and a deep enough understanding of operations that the transition to algebra is easy. From our perspective, none of the curricula we were working with treated computation sufficiently. Some focused on developing a repertoire of pencil/paper strategies; some designated focus algorithms; others focused primarily on learner-invented strategies; and some primarily made use of hands-on materials such as base blocks to teach the standard algorithms. None really pushed children to generalize or transition to algebra. In fact, when algebra was taught, it was usually seen as a separate strand characterized by analysis of patterns and functions.

To strengthen computation, we began to design minilessons with strings of related problems to develop deep number sense and a repertoire of strategies for mental arithmetic. Our goal in designing these was to encourage children to look at the numbers first before they decided on a strategy, and to have a deep enough sense of landmark numbers and operation that a toolbox of strategies could be used to calculate efficiently and elegantly—like mathematicians.

Over the years, to help our teachers develop vibrant math communities, rather than developing isolated hands-on activities, we built *sequences of investigations* to ensure progressive mathematics development. Several of the sequences we designed focused on developing mathematical modeling, for example, the open number line and the open array. Once these models are developed they can be used as powerful tools for thinking—for generalizing, proving, and even doing algebra.

For three years we ran a think tank on the emergence of algebra. Building on our initial work on number and operation, we field-tested sequences using the open number line to develop equations and strategies for solving for unknowns. We worked on encouraging children to develop conjectures and proofs. We helped teachers find the moments to push for generalization—to extend the work on number to algebra. Along the way we began to gain an understanding of how algebra might be taught in the elementary school.

This book is a culmination of that work. In it we describe our views on algebra and its development. We provide a "landscape of learning"—a trajectory of big ideas, strategies, and models for algebra—depicting the landmarks or milestones to be supported and celebrated. We tell the stories of many talented teachers and their students hard at work exploring and structuring their lived worlds algebraically.

ABOUT THIS BOOK

Chapter 1 describes and illustrates our beliefs about what algebra is and how it might be developed in the elementary years. We discuss it as *structuring*, but we ground it in the progression of strategies, the development of big ideas, and the emergence of modeling because we hold a constructivist view of learning. The mathematician Hans Freudenthal once commented that mathematics should be thought of as a human activity of "mathematizing"—not as a discipline of structures to be transmitted, discovered, or even constructed—but as schematizing, structuring, and modeling the world mathematically. This quote served as a framework for us as we studied and documented the emergence of algebra in many classrooms over the past five years.

Chapter 2 explains and presents a "landscape of learning" for algebra. For teachers to open up their teaching, they need to have a deep understanding of this landscape, of the strategies, big ideas, and models children construct, of the landmarks they pass as they journey toward the development of rich, dense structures characteristic of many algebraic relations.

Chapter 3 takes us back to the K–1 classroom to see the early part of the journey—young children beginning to structure the number system. Their structuring shifts from counting strategies and additive structuring to an early form of multiplicative structuring as they explore even and odd numbers. Their journey is facilitated by way of games, contexts, and investigations previously published (also by Heinemann) in our *Contexts for Learning* unit, *Beads and Shoes, Making Twos*.

Chapter 4 continues with the journey of structuring the number system as fourth graders explore geometric models of number, factoring, and the associative, commutative, and distributive properties for multiplication. Their journey is facilitated by the sequence of minilessons and investigations in the *Contexts for Learning* unit, *The Box Factory*. Related video of this chapter can be found in our professional development resource package (also published by Heinemann)—*Working with the Array*.

It is impossible to talk about mathematizing without talking about modeling. Chapter 5 introduces the power of the double number line as children in grades 2 and 3 explore equivalence and develop algebraic strategies such as substituting equivalent expressions and cancelling equal amounts when analyzing equations. In this chapter we also describe the importance of developing children's ability to treat an expression as an object that can be operated on (in contrast to a procedure) and we discuss the difference between this algebraic strategy and an arithmetic strategy. The investigations used by the teacher in this chapter can be found in *Trades, Jumps, and Stops*.

The focus of Chapter 6 is variation. In this chapter we include a brief history of the development of algebra and parallel it with the emergence of algebra ideas in children. We discuss the importance of introducing variables in ways that emphasize relations, not just simply as unknowns and share stories of fifth-grade children being introduced to symbolizing with variables as they investigate *The California Frog Jumping Contest*.

In Chapter 7 we turn to the development of integers, arguing that extending the number system to include integers means developing mental images of negative numbers. We describe why we consider chip and number line models to be insufficient and offer a beginning look at the net gain and loss contexts we are currently exploring.

Chapter 8 discusses combination charts and double number lines as powerful tools for comparing quantities and exploring systems of equations. Returning to the context of *The California Frog Jumping Contest* we see children engaged in determining lengths of unknown jumps by simplifying algebraic expressions.

In Chapter 9 we turn to the topic of minilessons: how to craft them and use them to ensure that the big ideas of algebra (such as equivalence and variation) are at the heart of the work.

Last, in Chapter 10, we focus on the topic of proof. We describe the difference between writing about what one did to solve a problem and the crafting of an argument for the mathematical community to read and comment on. How do we help children (and teachers) begin to see themselves as mathematicians, be willing to inquire, work at their mathematical edge, appreciate puzzlement? We open a window into classrooms engaged in the writing of proofs and invite you to examine proof-making along with the teachers and children we describe.

Like all human beings, mathematicians find ways to make sense of their reality. They symbolize relationships, quantify them, and prove them to others. In the process they develop a rich network of relations—a powerful lens that they use in structuring. For teachers to engage children in this process, they must understand and appreciate the nature of mathematics. They must be willing to investigate and inquire—and to derive enjoyment from doing so. The book you hold is primarily about that—how teachers and children come to structure their own lived worlds mathematically, their journeys as they pursue the hard work of constructing big ideas, strategies, and mathematical models in the collaborative community of the classroom.

1 | ALGEBRA: STRUCTURES OR STRUCTURING?

When I have clarified and exhausted a subject, then I turn away from it, in order to go into darkness again. The never-satisfied man is so strange, for he completes a structure not in order to dwell in it peacefully, but in order to begin another.

—Karl Friedrich Gauss, Letter to Bolyai (1808)

Nothing is more important than to see the sources of invention, which are in my opinion more interesting than the inventions themselves.

—G. W. Leibniz (1646–1716)

What is algebra? How does it develop? Because algebra includes many things—generalizing beyond specific instances, describing and representing patterns and functions, building equations and expressions using symbolic representations with integers and variables, manipulating symbols to solve for unknowns—there has been a spirited debate as researchers have tried to define the topic, specifically as it relates to teaching this strand of mathematics in elementary and middle school.

Some researchers have argued that algebra in the elementary school should be thought of as the construction of algebraic "big ideas" growing out of generalized arithmetic (Schifter, Russell, and Bastable 2006; Carpenter, Franke, and Levi 2003). Others (Driscoll 1999) describe algebra as a type of reasoning in which one investigates the relationships between specific cases and possible generalizations and develops algebraic "habits of mind"—ways of thinking about algebraic questions. Whereas early work on algebra in the sixties' "new math" movement focused on examining algebraic structures (Wirtz, cited in Goldenburg and Shteingold 2008), more recent work has examined the process of "algebrafying" (for example, systematically symbolizing generalizations) (Kaput, Carraher, and Blanton 2008).

Over the years, our own participation in this discussion was not directed toward defining algebra as much as it was toward studying its emergence. Our questions have been: What might the development of algebra look like in the elementary grades? What are some of the critical big ideas and strategies young children construct that might serve as important landmarks for teachers to notice, develop, and celebrate? What causes some of the misconceptions and challenges that develop? How might realistic contexts and representational models—a double number line, combination charts, and the ratio table, for example—support the development of algebra? As we worked in classrooms attempting to answer these questions, our own working definition of early algebra began to emerge.

TEACHING AND LEARNING IN THE ALGEBRA CLASSROOM

It's a beautiful crisp fall day in New York City. Bill is leading a Math in the City (MitC) professional development session for twelve elementary teachers. The group has enthusiastically agreed to come together for four days over the course of the semester to deepen their own understanding of algebra and to study how the topic might be taught in the elementary grades. In a typical session the participants explore and discuss a mathematical investigation, choose an inherent idea, and craft a context in which to explore this idea in a fifth-grade classroom in the host school. Then one of the participants teaches the lesson while the others observe. Finally everyone reflects on the teaching and learning that has taken place. Today Bill is focusing on algebraic structures in the number system.

"How many factors does the number 1 have?" Bill begins.

The question is easy. "Just one," several teachers exclaim immediately.

"And the numbers 2 and 3 each have two. What about the number 4?"

"That has three," Camille, a seasoned fourth-grade teacher, replies quickly. "1, 2, and 4 are all factors of 4."

"And the number 6 has four factors, right, 1, 2, 3, and 6? Okay. So let's investigate this. Work with a partner and sort numbers by the number of factors they have. Let's see if we can find any interesting relationships."

Camille and Karen, working with one number at a time, generate a table to show their results (see Figure 1.1). When they reach 13, they begin to discuss what they have noticed.

"Thirteen's a prime number. All the primes are going to have just two factors," Camille says, filling in the chart's cells for 13. Then she whispers, "You know, this is embarrassing; I knew 1 wasn't a prime number because a teacher told me that once and I accepted it. But my definition of primes was a number that had factors of 1 and itself only. I never quite understood why 1 wasn't a prime because it fit my definition, but in our list I see now why it isn't. It's by itself. All the primes are going to have two factors. The number 1 only has one factor."

Karen nods. "I think I'm seeing another pattern. It seems that if we double a prime number, we have to add two to the number of factors. See, 3 is a prime number, right? And it has two factors. If we double 3, we get 6, and that number has four factors. The number 5 has two factors, and 10 has four. I think 14 will have four factors, too. Yep—1, 7, 2, and 14."

Camille is intrigued. "That's interesting." She ponders a minute and then realizes what is happening. "Oh, I get it. We are adding two more factors. First it was 1 and 7, but when we doubled the number, the new product and the number 2 became factors, too. That should always work."

"Yeah, that makes sense, but why didn't it work with the number 2? The number 4 has three factors, not four." She thinks a moment. "Maybe because 2 was already a factor? Hey, maybe odds and evens have something to do with it?" *Sometimes finding an example that disproves a conjecture can prompt a new insight.* "Hey, the rest of the prime numbers are odd! I just realized . . . of course, they have to be, because to be even 2 would have to be a factor!"

Delighted, Karen writes on their chart, "All primes except 2 are odd and when you double an odd prime you get two more factors—so plus two."

Meanwhile, across the room George and Maria have created a horizontal chart showing the number of factors as numbers increase by one (see Figure 1.2).

George, chunking the numbers into overlapping triads, notes a pattern. "This is interesting. I see a 232, then a 242, but then a 243 and a 342. The numbers are always larger in the middle."

"Yes, I see that but I don't see what the pattern really is. I mean I don't think we can use it to predict it."

Number	Factors	Number of Factors
1	1	1
2	1, 2	2
3	1, 3	2
4	1, 2, 4	3
5	1, 5	2
6	1, 2, 3, 6	4
7	1, 7	2
8	1, 2, 4, 8	4
9	1, 3, 9	3
10	1, 2, 5, 10	4
11	1, 11	2
12	1, 2, 3, 4, 6, 12	6
13		

FIGURE 1.1
*Camille and
Karen's Table*

Bill has been listening to this conversation and joins the pair. "George, are you thinking about going from one number to the next?"

FIGURE 1.2
George and Maria's Chart

1	2	3	4	5	6	7	8	9	10	11
1	2	2	3	2	4	2	4	3	4	2

"Yes, isn't that what you do when you search for patterns? You asked us to look for patterns, right?"

"Well, I didn't exactly ask you to search for patterns. I think I said something about relationships you might notice when you sort numbers by their number of factors." Maria still looks puzzled, so Bill attempts to reframe the investigation. "Try thinking about this. Factors depend upon multiplication, right? Maybe it would help to sort the numbers into piles with the same number of factors and think about how they are related in terms of multiplication."

Maria begins to formulate a new direction for their inquiry. "You mean like, six is two times three and ten is two times five? They are both double odds."

"That's an interesting idea. Keep going and I'll check back with you in a bit."

In many classrooms algebra is taught as recognizing and extending patterns, and George and Maria initially try to do that by looking for sequential patterns. While certainly a part of early algebra, this notion is not sufficiently general. Bill has deliberately chosen an investigation in which sequential (or additive) structuring will not make sense. Numbers can be structured in many ways: in a sequential (plus one) order, additively, multiplicatively, exponentially, and so on. The paradigmatic shift from additive to multiplicative structuring took centuries!

Bill returns to Camille and Karen to see whether they are making progress with their "plus two" idea when they double an odd prime. Camille begins generalizing her observation. "Hey, maybe it's not just primes! Maybe it's all odd numbers. Nine has three factors: 1, 3, and 9 . . . oh no, 18 is the double but it has six factors: 1, 18, 2, 9, 3, and 6. If it had been plus two, it should have had only five factors." She lays out the original pairs for 9 and then compares them with the pairs for 18 (see Figure 1.3).

FIGURE 1.3
*Camille's Organization
of Factors of 18*

1, 9	1, 18
3	2, 9
	3, 6
Factors of 9	Factors of 18

"It's like they reorganized themselves," Karen says. "The 18 went with the 1; the 9 slid down to go with the 2; and 6 appeared for the 3. The number doubled, and the number of factors doubled. Let's try another one. Let's do 75. I want to look at a bigger number and see what happens." Together they make a factor tree and then chart pairs of factors for the numbers 75 and 150 (see Figure 1.4).

"So when you double an odd number, the number of factors doubles, too? Will this always happen?" Karen asks.

"The new factors are double all of the original ones, too," Camille observes. "It must always work. Let's see what happens when we double even numbers. Let's try 6. It has factors of 1, 2, 3, and 6. Twelve has factors of 1, 12, 2, 6, 3, and 4. Shoot—only 6 factors, not 8. It doesn't work. What's going on here?"

Traditionally, students have algebraic structures explained to them. For example, they may have been told that the natural numbers can be sorted into evens or odds, or into primes and composites. They are also taught that every number has a unique factorization into products of prime powers and that least common multiples are helpful for adding fractions. Although they may understand these structures, their understanding may be disconnected, comprising separate unrelated categories, and may be directly linked to specific actions (such as even numbers can be divided by 2, or making factor trees to produce factors and factor pairs). Richer, "dense" structures are derived by exploring and setting up a variety of relations, often using a larger set of operations.

1×75	
3×25	
5×5	
1, 75	1, 150
3, 25	2, 75
5, 15	3, 50
	5, 30
	6, 25
	10, 15
Factor pairs of 75	Factor pairs of 150

FIGURE 1.4
*Camille and Karen's
Factors of 75*

By asking these teachers to sort numbers according to their factors, rather than defining and presenting an already formed structure, Bill gives them the opportunity to structure the number system in new ways that will provide a better understanding of multiplicative relationships. As they inquire they are setting up relations and analyzing them, examining relationships, and generalizing—they are doing mathematics. Algebra is often described as an act of generalizing, which while no doubt true, skirts a key piece of development—the structuring of the objects into part/whole relations. It is not possible to generalize if one has not first structured the objects at hand by setting up correspondences between them, examining how one is transformed into another, and how they are related to the number system as a whole. It is precisely these actions that potentially result in the construction of a denser set of interconnected relationships characteristic of a broader and deeper understanding (Piaget 1977).

Initially, Karen and Camille's structuring is like George and Maria's—additive rather than multiplicative. At first they explore sequentially, one number at a time in a plus one fashion (1, 2, 3, 4 . . .). The increase they find as they double the primes they originally construe as plus two. However as they work, they shift to a doubling strategy—an emergent multiplicative form of structuring (if the operation is seen as $2n$ rather than $n + n$)—and begin to discuss the resulting increase in factors as a double as well. Notably absent from their strategy, however, is the examination of how the prime factors are acting on one another multiplicatively.

Across the room, Marion and Marco are structuring multiplicatively. They start by sorting numbers by number of factors (see Figure 1.5).

They quickly realize that all primes are in the two-factor column and then begin to notice relationships across the columns.

"Hey, look at the three-factor column," Marion says excitedly. "They are the squares of the primes!"

"Awesome!" Marco grins. Then he looks perplexed. "But why is that happening?" He ponders a moment. "Oh, I get it! The primes only have two factors, so when you square a prime number you only get one more new factor, and it's the product of the squaring. See? The factors of 2 are 1 and 2, and when you square 2, you get 4. So now there are three factors, 1, 2, and 4."

"What if we cube a prime? Like 2 times 2 times 2. That's 8, and 8 is in the four-factor column!"

"And 3 times 3 times 3 is 27. Are you saying 27 should be in the four-factor column, too?" Marco asks.

"Yes. Because each time you multiply the prime factor, you just get one new factor. Like you said before—it's the multiple. The factors of 27 are 1, 3, 9, and 27. Before, for 3 squared, we had 1, 3, and 9. When we cubed 3, we added 27 as a factor. I think I'm about to be brilliant," Marion giggles. "When multiplying prime numbers by themselves, the number of factors is one more than the exponent: 3 squared has three factors; 3 to the third power has four factors; 3 to the fourth power has five factors, and so on."

"Yep, you are awesome!" Marco agrees. "Let's redo our chart to show those results." They eliminate the numbers that don't fit their pattern and begin adding numbers that do (see Figure 1.6).

"So now our table looks nice, and we are sure about this piece, but what about the numbers we removed?"

Although they have completed one structure regarding the exponentiation of primes and are investigating the multiplicative nature of this process, they

One Factor	Two Factors	Three Factors	Four Factors	Five Factors	Six Factors	Seven Factors	Eight Factors
1	2	4	6	16	12		24
	3	9	8		18		
	5	25	10		20		
	7		14				
	11		15				
	13		21				
	17		22				
	19						
	23						

FIGURE 1.5   *Marion and Marco's Chart*

One Factor	Two Factors (Primes)	Three Factors (Primes Squared)	Four Factors (Primes Cubed)	Five Factors (Primes to the Fourth Power)	Six Factors (Primes to the Fifth Power)
1	2	4	8	16	32
	3	9	27	81	243
	5	25	125	625	3125
	7	49			
	11	121			
	13	169			

FIGURE 1.6   *Marion and Marco Add to Their Chart*

are still thinking about the growth process of the factors as additive (the exponent plus one). An important unanswered question surrounds the remaining nonprime powers. Susannah and Malika, sitting nearby, are entertaining this same question.

"I just don't get how we could know how many factors these other numbers have," Susannah exclaims with exasperation.

"I know. This is so hard!" Malika agrees.

Bill tries to be supportive. "This is a pretty complex inquiry, isn't it? But that's what also makes it fun. What are you confident of so far? Sometimes it helps to step back and reflect on what has worked thus far and then plan your new direction."

So often teachers feel they should give answers or provide procedures to learners. The problem with doing so, however, is that those same learners then come to depend on the teacher's hints and never develop the tenacity to work through the important struggles. Even more to the point, they never learn to appreciate the fun of puzzlement and the exhilaration that comes with their own breakthroughs. Keith Devlin (2003) once said, "When I'm working on a problem it's like climbing a mountain. Sometimes I can't even see where I'm going. It is one foot in front of another. And then I reach a point where all of a sudden the vistas open up and I can go down easily for a while, only to eventually reach another climb." As teachers, our goal is to build the learner's capacity to make the climb. To that end, when we confer with learners we need to focus on developing the mathematician rather than fixing the mathematics. Every action we take should develop the novice mathematicians in front of us.

"We know that prime numbers squared produce one more factor," Susannah summarizes.

Malika extends the idea by symbolizing it. "So what we mean is the exponent matters, because the number of factors is just one more. Like 3 to the fifth power has six factors. The exponent is 5 so there are five plus one factors."

"So where are you stuck?" Bill asks.

"We have other four-factor numbers, too, like 6, 10, 14, 15, 21 . . . lots of them, and we can't figure out why, or to put it better, how we could know ahead that they would be four-factor numbers."

Bill paraphrases: "So you're saying you are confident that primes are important, and you know that multiplying primes by themselves produces one more factor. Have you looked at what happens when you multiply two primes that are different? Like 2 times 3, instead of 2 times 2? That might be something you could pursue."

"Look—2 times 3 is 6 and that is in the four-factor column," Malika says, giving Susannah an enthusiastic nudge. "Ten is there, too, and that is 2 times 5."

Susannah offers a new conjecture. "So if each of the primes has two factors, and these have four, maybe we just add the number of factors of the primes—two plus two equals four." She is looking at the composition additively, just as Camille and Karen have done.

"Yes. That seems to work. Let's try a prime number times a nonprime, like 3 times 4; 3 is prime, but 4 is not. How many factors does the number 12 have?"

Malika's question is an important one: Her remaining additive structuring is about to prove insufficient.

"Twelve is a six-factor number. But 4 is a three-factor number, and 3 is a prime, so it has two factors. Twelve should be a five-factor number, not six! Two plus three equals five!" Susannah replies.

Malika is puzzled but intrigued and determined. "Maybe we should multiply the number of factors. Two times three equals six. If we use what we know about the primes, we would know how many factors each number has. Like 2 squared times 3 squared. That's 4 times 9. Thirty-six, right? We know from our earlier rule that if the power is 2, it has three factors. So both of these numbers have three factors. Let's try multiplication again. Three times three equals nine. Thirty-six should have nine factors."

Susannah has been listing the factors of 36 as Malika talks: 1, 2, 3, 4, 6, 9, 12, 18, and 36. She counts them and exclaims, "It does, it does! Hot dog! We're good!"

"So now we have to figure out why! Why do we multiply? We need to prove it."

"We will. We will. We're on a roll! Let's keep going. I think maybe it has to do with the number of ways we can multiply the factors to make new factors—the number of combinations that can be made."

WHAT IS REVEALED

The mathematician Hans Freudenthal wrote, "Mathematics should be thought of as a human activity of mathematizing—not as a discipline of structures to be transmitted, discovered, or even constructed—but as schematizing, structuring, and modeling the world mathematically." His point was that development should be emphasized and fostered.

Rather than trying to explain the algebraic structures related to prime factorization, Bill encourages these teachers to structure the number system according to numbers of factors in their own way. As they work, he challenges and supports them to develop more powerful ways of structuring. This developmental approach is also in contrast to discovery learning approaches, in which learners complete an activity designed to produce the same "aha" for everyone at the end, as they discover (uncover) the structure that was the teacher's intent. In both transmission and discovery models of learning, preformed algebraic structures are the focus. The teacher attempts either to transmit the algebraic structure or help learners discover it.

The focus of Bill's work is the *development of structuring—the progressive building* of algebraic structures and the mathematical *development of the learners*. Bill wants to help his learners progressively develop richer ways to structure the number system. Rather than presenting them with a *prestruc-*

tured world organized by the past activity of mathematicians throughout history, he presents them with a mathematical *world to be structured.* In so doing, he invites them to do mathematics, to find ways of organizing and categorizing number by examining common features, similarities, and relations as a way toward generalizing. He invites them to participate in the working world of the mathematician.

It is human to seek and build relations. The mind cannot process the multitude of stimuli in our surroundings and make meaning of them without developing a network of relations. Throughout our history, we have found ways to structure our lived worlds. We categorize, seriate, and compartmentalize. We examine, evaluate, and compare; we make connections and set up correspondences. We develop systems and describe how the parts of the systems are related to the whole. We have even created whole disciplines of knowledge comprising the relations we build between ideas. As students engage in structuring, the relationships they use and understand are transformed as they shift their cognitive lenses. Their structures become denser—the network of relations they build becomes richer. *When invited to engage in structuring, they build cohesive structures with many pathways and interconnections.* In contrast, when presented with preformed structures, students often process them as isolated bits, associated only with the activity or topic by which they were introduced.

While investigating how numbers are sorted according to number of factors, these teachers are building increasingly richer, denser structures. Marion and Marco have uncovered the multiplicative relationship that squares of primes are three-factor numbers. Their subsequent realization that this extends to higher powers of primes (e.g., cubes of primes are four-factor numbers) enriches their structure. Camille and Karen initially are thinking only about additive relationships: If you double an odd number, you get two more factors. However, this additive structuring allows them to deepen their understanding of prime numbers and the patterns created in the production of factor pairs. In their investigation of the factors of 36, Susannah and Malika are structuring the number of factors multiplicatively. As they explore unique factorization and the multiplicative relations of the factors, a robust algebraic structure emerges.

The participants in this community are walking the edge between additive and multiplicative structuring. Camille and Karen have noticed that if you double an odd three-factor number, then the number of factors doubles. But is this doubling being thought of as $3 + 3$ (additive) or as 2×3 (multiplicative)? They have organized their factors according to factor pairs, a procedure they have been teaching their students to use. Marion and Marco are examining the multiplicative aspects of exponentiation, but they are still describing the growing nature of the number of factors in an additive way. Susannah and Malika have found cases in which if you multiply two three-factor numbers you get a nine-factor number, not a six-factor number, so they have found a multiplicative relationship. But is this a generalizable idea for them? Do they understand why?

Structures become denser as students add relations: a variety of ways to symbolize equivalent relations, new observations that occur from similar ways of structuring, or new insights that occur from a shift in operation (for example, from additive to multiplicative). At this moment, how can Bill support all the members of this learning community to build denser algebraic structures? Can learners really construct powerful structures on their own that took humans centuries to invent? How can we, as teachers, facilitate such mathematical development?

Professional mathematicians share their ideas with one another through publication and conferences. They read one another's proofs, comment on them, and discuss ideas, strategies, and insights. Isaac Newton once wrote to Hooke, "If I have seen further it is by standing on the shoulders of giants." He was expressing his gratitude to the many talented scientists before him. By examining and comparing different or even contradictory ideas and working to confirm and extend ideas, mathematicians connect, enrich, and extend the structures of their discipline. If this process is knowledge-generative in the professional community of mathematicians, might it also be growth-producing in the classroom?

BACK TO THE WORKSHOP

Bill asks the participants to make and display posters of their work thus far. He then has everyone examine everyone else's work and place sticky notes with comments and questions on the posters. He has three purposes in mind. First, as participants make their posters they will have to consider what information is important to communicate and how they will justify their ideas to others in their mathematical community. Second, he hopes the experience of reading others' mathematics will challenge them to examine their own mathematical arguments and insights. Third, by encouraging the participants to reflect on and discuss the various ways of structuring and the relationships that have been noticed thus far, Bill is preparing them for a subsequent "math congress" in which he will facilitate a discussion on the power of shifting from additive to multiplicative structuring. He begins the congress by asking Camille and Karen to explain how they have represented the six factors of 18 on their poster.

Camille begins. "Well, we noticed that the number of factors doubles, so we put them in pairs. See, 1 and 18 go together, 2 and 9 go together, and 3 and 6 go together."

"That's right," Karen chimes in. "We put them into factor pairs. Each pair goes together because they multiply to give 18. We've been teaching our kids how to make factor pairs just this very week."

"And what do the pairs have to do with the doubling you noticed?" Bill asks.

"We think that when you have pairs you have a doubling. That's why there is an even number of factors."

George and Maria, who have proceeded quite differently, look perplexed. George says, "We're confused. Nine has factor pairs, too. How are these doubling in your diagram?"

Karen writes the factors of 9 in a column next to the factor pairs. She writes 1, 3, 9, and then continues with her explanation. "See, each factor of nine becomes part of one of the factor pairs of 18. So since we have factor pairs now, that's twice as many. That's why the number of factors doubles."

Murmurs of understanding can be heard among the group.

Next, Bill asks Susannah and Malika to share how they represented the factors. They have not organized the factors in pairs as Camille and Karen have done but instead have doubled each of the factors of nine (see Figure 1.7).

Bill says to Susannah, "Tell us why you and Malika decided to organize them in that way."

"Well, since we were doubling the number, we decided to double the factors, too. We did that because when we tried to find the factors for 36, we wanted to know why there were 9 factors."

George says, "I guess I don't see the point. Don't you want to have factor pairs? That's how you can keep track of the factors. Camille and Karen just convinced me of that."

"That's true," says Malika, "but where we really got the idea was when we did 36. At first we couldn't figure out why 36 had 9 factors. We thought it should have 6 because 36 equals 4 times 9 and 4 is a three-factor number and 9 is a three-factor number. We thought the three should double and be six, too, just like Camille said. But it didn't work. We really wanted to know why you got nine factors there instead of six. We knew 3 times 3 was 9, so we thought maybe we should look at this with multiplication."

Bill suggests, "Malika, why don't you show us what you mean by extending your chart for 18 to include 36. Take a different-color marker so we can keep track of the new factors you add."

This shift from additive structuring to multiplicative structuring is a big idea—one that requires cognitive reorganization on the part of learners—and they will need time to consider this shift in perspective.

Malika draws the representation shown below and then elaborates. "I'm multiplying the first column of factors by 4. This way we get a whole new column of factors [*pointing to the third column*]. We add 4 times 1, then 4 times 3, and then 4 times 9" (see Figure 1.8).

FIGURE 1.7
Comparing Two Ways of Structuring the Factors of 18

1	1, 18	1, 2
3	2, 9	3, 6
9	3, 6	9, 18
Factors of 9	Factor pairs of 18 (Camille and Karen)	Factors of 18 (Susannah and Malika)

*Manipulation of numbers to produce an answer can seem like a magic trick
to learners if they haven't constructed the implicit relations for themselves. The
importance of this construction cannot be overemphasized, because it is precisely
what enables learners to generalize. To that end, Bill asks the crucial question.*

"How do you know these are new factors? That's an important question. Take some time to reflect on this, and then talk to the person next to you about your thoughts."

After a few moments he resumes whole-group discussion. "Malika, what are your thoughts about this?"

"Well, they are new factors, if you check them."

"Okay, but do you think there is some way you could know without checking that they have to be new factors?"

Susannah says tentatively, "There must be some way. I think it has to do with the fact that 4 is a factor of 36 but not a factor of 9."

Camille has an insight. "Wait a minute—yes, that's it, Susannah. I think I see it. If these new numbers in the third column are divisible by four . . . I mean, these new factors are divisible by 4, but none of the factors of 18 were divisible by 4, so they have to be new!"

Karen shakes her head in amazement. "This is so different from what we were doing. We were looking at factor pairs that multiply to 18. But now you are listing in each column factors with more and more twos! First no twos, then one two, then a four, which is two twos—that's why I think you are getting 3 times 3 factors. You are getting the factors of 9 on the side of the array, 1, 3, and 9, and the factors of 4 across the top, 1, 2, and 4. It's like the array model that we use for multiplication but with factors, and it's showing all of the possibilities! We don't have factor pairs any more. We have rows and columns!"

Maria adds, "It *is* kind of like the array model for multiplication! The top row is the factors of 4 and the left column has the factors of 9."

*For Maria, the 3 × 3 array suggests that the product of 3 × 3 is what is
important in counting factors.*

*Karen and Camille's thinking has been transformed. They are now looking at
the multiplicative relationships between the factors of 36. The comparison of the
two different ways of representing factors has helped them (and other members of
the class) reconsider their original structuring of the factors of 18. Instead of
organizing their thinking around the factor pairs that multiply to 18, they are
now sorting the factors according to divisibility relationships among the factors*

1,	2,	**4**
3,	6,	**12**
9,	18,	**36**

FIGURE 1.8
*Malika Adds Factors of
36 to Factors of 18*

themselves (not simply as factors of 36). So for the example of 36, the factors 2, 6, 18 are distinguished from the factors 4, 12, 36 because the latter three are divisible by 4, while the previous three are divisible by 2 but not by 4. This is an important shift in thinking for several reasons. First, for many participants the factors of 36 divided only into even and odd factors, but now a new divisibility relation is on the table (the odd factors, the factors divisible by 2 but not 4, and the factors divisible by 4). Second, this shift makes it possible to figure out why the number of factors of 36 is the product of the number of factors of 4 and the number of factors of 9, and this type of structuring makes generalizing possible. Finally, it will open doors to other ways to structure the number system, including properties of unique factorization and exponentiation.

VISITING A FIFTH-GRADE CLASSROOM

The teachers in Bill's workshop, by structuring the number system, have developed some important algebraic structures. But what about children in elementary school? What bearing does this work have on their instruction? Wouldn't such open-ended inquiry be too difficult and perhaps even confusing? How far would their structuring take them? To find out, the group discusses what investigation they would like to see a fifth-grade class tackle that afternoon.

Camille's fifth graders have been studying factoring by making factor trees and creating factor rainbows for numbers, in which the two factors whose product is the original number are connected by an arc. (For example, if the factors of 12 are listed horizontally in ascending order: 1, 2, 3, 4, 6, 12, a "rainbow" of three arcs would be formed, the smallest over the 3 and 4, the next over the 2 and 6, and the third over 1 and 12.)

"I think opening up the activity to a fuller inquiry is what made this so exciting and rich for me," Camille confesses. "Making factor rainbows seems so closed and trivial now. I mean, why would kids want to do that anyway? They're just doing it because I asked them to. But what are they really learning?"

"What do you think they would do with this investigation?" Bill asks.

"I don't know, but I think they could sort the numbers, and it would be interesting to find out what they notice."

The participants head for Camille's class. She introduces the investigation to her fifth graders in a manner similar to that used by Bill in the morning. Being asked to structure numbers in this way is a new experience for most of the students, yet they seem to have a way to start. They quickly set to work in pairs. The teachers move through the room, listening and observing.

"The odd numbers are the two-factor numbers. See, 3, 5, 7, and so on. Odd numbers will have two factors I bet," Sam says to his partner, Emilio.

Emilio tries 9. "No. Doesn't work. See, 9 has 3 factors: 1, 3, and 9. Look at my list. I think it's prime numbers that have two factors."

"Oh, yeah. I think that's the answer we're supposed to get. It's about primes and composites. Primes have two factors and the composites have three or four."

Camille, who is standing nearby, joins in. "What can you say about three-factor numbers? And what can you say about four-factor numbers? Keep going."

Sam looks confused. "I don't get what you mean." Usually in school when you find the teacher's answer you're done. Hadn't he found it? What could she possibly mean by "keep going"?

This is a new role for Camille as well. Usually she would have congratulated the pair and tried to find something else for them to work on while the other children finished. Now she's trying to model Bill's line of questioning, all the time wondering, "How can I get them to be as excited as I was this morning?" *Hook them on an inquiry*, she thinks as she forges on. "Well, you say the two-factor numbers are primes. What's true about a three-factor number that makes them different from four-factor numbers?"

Sam offers tentatively, "Well they are odd maybe . . . except 4. Maybe the four-factor numbers are even. . . ."

"Interesting thought. See if you are right. I'll be back in a little while to check with you and you can let me know what you find."

Sam and Emilio resume investigating, and Camille grins with satisfaction as she leaves them and goes to see what other pairs of students are doing. She did it! Sam and Emilio are intrigued. This is fun. Her kids are busy investigating. Amazing! It's true that other than noticing the primes, most students are sorting by evens and odds, and some are spending a lot of time developing their lists of numbers of factors. She notes with interest that only two students are using the factor rainbows she's had them practice making all week.

The following is a typical conversation.

"You have 24 here under eight-factor numbers. How did you know that?"

"1, 2, 3, 4, 6, 8, 12, 24."

"That's great! How do you keep track of all of those?"

"I think about the multiplication tables."

"Is there any other way to think about all of these factors?"

"I just multiply in my head and I get them."

After about half an hour, the class has collected quite a bit of data, and most students have recognized the primes as being the two-factor numbers. To push their thinking Camille calls the class together for a short congress. She draws a chart on the board and then asks her students to provide data for it from their work. Together they produced the chart shown in Figure 1.9.

After providing some time for partner talk about the relationships students see on the chart, Camille begins a whole-class discussion.

"Sam, you and Emilio were talking about primes. Tell us about your idea."

"They are the two-factor numbers. It kind of has to be true. Prime numbers don't have any other factors."

"Okay. So we have a generalization for two-factor numbers. What are your ideas about the three-factor numbers?" At first no one responds. Then slowly a few hands go up. Camille calls on Anthony.

He offers an interesting observation, although a bit tentatively. "I think maybe the numbers in the third column are the square roots of the numbers. Four is the square root of 2 and 9 is the square root of 3, and like that. They are square root numbers."

In the past, Camille would have immediately focused on Anthony's misuse of terminology. But correcting him at this moment may cause him to shut down, just as he's beginning to shift the class discussion to multiplicative structuring. Therefore, she probes his thinking. "Tell us more about what you are noticing."

"Well, 2 times 2 is 4, 3 times 3 is 9, 5 times 5 is 25. So these are square root numbers. And they are the square root numbers of the primes."

Work with the developing mathematicians, don't just try to fix the mathematics, Camille thinks. She can help Anthony with the correct vocabulary later. Right now the class needs time to consider the relations he is describing. "Talk with your math partner about this for a minute. What do you think about what Anthony just said?"

Within a few minutes the room is a buzz of voices. "He's right. Look! Multiply the prime numbers! Two times 2 makes 4, 3 times 3 makes 9, 5 times 5 makes 25! Those are all square numbers. The three-factor numbers are square numbers of the primes!"

"And if you square a three-factor number you get a five-factor number!"

"And if you double the prime numbers you get a four-factor number. Look at the chart. See how the four-factor numbers go? They go 6, 10, then 14. These are four-factor numbers and they are two times the odd two-factor numbers, 2 times 3, 2 times 5, and 2 times 7."

One-Factor Numbers	Two-Factor Numbers	Three-Factor Numbers	Four-Factor Numbers	Five-Factor Numbers	Six-Factor Numbers
1	2	4	6	16	12
	3	9	10	81	32
	5	25	14		
	7				
	11				
	13				

FIGURE 1.9 *Chart from Camille's Class Congress*

"Hey, multiply any two different prime numbers and you get a four-factor number!"

Camille is elated. "Wow! I'm hearing so many neat ideas! So what do you think the next four-factor number will be? Amirah?"

"I think 18 because that's 2 times 9."

Sam is looking at the chart he and Emilio have made. "No, it can't be 18. That's a six-factor number. It's 15 I think. And that's 3 times 5. Two primes."

Emilio shakes his head in disagreement. "But 15 is an odd number. I don't think odd numbers can be four-factor numbers."

"Yes, it is 15!" Amirah says, agreeing with Sam. "1, 3, 5, and 15. It does have four factors. Both odd and even numbers work. But look, the first number in every column is an even number!"

"So now we have some good questions to investigate. Let's go back to work and continue investigating." Camille knows they need time to explore further, but now they have a multitude of observations and questions that can direct their explorations.

WHAT IS REVEALED

"I can't believe your kids! That was just amazing what they came up with in such a short time!" Karen says to Camille, giving her a hug, when the teachers convene in a small room at the school to discuss their observations. Everyone agrees.

Then Bill focuses the group on the development of structuring and the role of the teacher. "What development did you see, and what did Camille do that supported it?"

"One of the things I found so interesting," George begins, "is how hard it is to think multiplicatively. I know how hard it was for me this morning, and when I walked around watching what the kids were doing at first, they were just doing one number at a time."

"George is right," Marion chimes in. "I did see lots of interesting things. One kid was making a bar graph for each number one at a time, showing how many factors each had. A lot of them were talking about even and odd numbers, and most noticed prime numbers were two-factor numbers. But nobody started multiplying until Anthony started talking about squares."

"I think your chart helped them notice the squaring," Marco says. "The columns were organized so that the squares of the primes were right next to the primes. It helped the kids notice the relationships."

"That's important, Marco," Bill agrees. "Did you notice that I also used them this morning when I juxtaposed Camille and Karen's factor pairing representation next to Malika and Susannah's array model?"

"And it helped me," Camille remembered. "What I mostly loved this afternoon though was that I really felt my students were growing as mathematicians. I felt the focus was really on their development. It felt so genuine. Halfway through the session I realized I had been doing all of this factor rainbow practice and no one was even using it! Instead they were using what they knew about multiplication. It made me wonder about so many of the cute things we do that we think will help kids but that are nothing but trivial activities. I like that my kids and all of us this morning were really finding so many new relationships among numbers. I don't think I'll ever look at numbers in the same way again!"

SUMMING UP

The workshop participants and the students in Camille's fifth-grade class have been structuring the number system. When algebra is understood as structuring, rather than as the transmission and examination of pre-formed algebraic rules, teaching and learning become seen as interwoven processes related to the development of the learner as a mathematician. Karl Friedrich Gauss said, "When I have clarified and exhausted a subject, then I turn away from it, in order to go into darkness again. The never-satisfied man is so strange, for he completes a structure not in order to dwell in it peacefully, but in order to begin another." Rather than "fixing" the mathematics in student productions as has often traditionally been the practice, teachers (like Bill and Camille) are finding ways to support and challenge learners to engage them in the making of mathematics. They are inviting them to derive enjoyment in the act of making algebraic structures and in the "going into the darkness again" in order to "begin another." In so doing, learners are developing denser, richer structures constituting many relations—powerful algebraic networks that will serve them in the years ahead.

The multiplicative structures traditionally taught in Camille's school in fifth grade included finding factor trees, factor pairs, unique factorizations, greatest common factors, least common multiples, and more. But the curriculum designers included them because they wanted students to apply these rules and procedures elsewhere, most likely when adding and subtracting fractions, not because they wanted students to engage in structuring the number system. Because these structures had been taught in isolation, when actually placed in a problematic situation requiring structuring as part of the sense making, the students (and the teachers) did not use them.

G. W. Leibniz observed, "Nothing is more important than to see the sources of invention, which are in my opinion more interesting than the inventions themselves." By opening up the task and inviting students to find ways to structure the number system, the sources of invention and the development of structuring became apparent. The first shift from additive to emergent multiplicative structuring was Karen and Camille's investigation of factors when doubling primes. This emergent multiplicative strategy allowed participants to move beyond thinking about factor pairs and see the relations embedded in the array structure introduced by Susannah and Malika. This richer multiplicative structuring led to an exploration of exponentiation including the roles of powers of primes in unique factorization. Slowly a more robust algebraic multiplicative form of structuring began to emerge. It was the carefully constructed context, use of representations, and questioning by Bill and Camille that supported this development. As the community engaged in powerful talk, they found ways to build on one another's ideas and generate more elegant structures.

2 | THE LANDSCAPE OF LEARNING

Cognition does not start with concepts, but rather the other way around: concepts are the results of cognitive processes. . . . How often haven't I been disappointed by mathematicians interested in education who narrowed mathematizing to its vertical component, as well as by educationalists turning to mathematics instruction who restricted it to the horizontal one.

—*Hans Freudenthal, China Lectures (1905–1990)*

Mathematics is not a careful march down a well-cleared highway, but a journey into a strange wilderness, where the explorers often get lost.

—*W. S. Anglin (1992)*

COMPARING CURRICULUM FRAMEWORKS

Historically, curriculum designers did not use a developmental framework when they designed texts, nor did they see mathematics (or algebra) as mathematizing—as structuring one's lived world mathematically. Instead, algebra was seen as the accumulated content of a mathematics strand comprising preformed algebraic structures, forms of symbolization, and generalized procedures for solving for unknowns. Teaching and learning objectives were determined by formulating this content into skills and concepts and placing them along a line (Gagne 1965; Bloom et al. 1971). For example, simplistic notions of patterns and solving missing addend problems were considered developmentally appropriate for early childhood as a way to begin the algebra strand. Later, in the middle elementary grades, learners were introduced to the commutative, associative, and distributive properties and to some of the structures in the number system (such as odd and even numbers, factors, and multiples). In middle school they studied proportional reasoning, integers, variables, and procedures for solving for unknowns in simultaneous equations. Development was considered but only in relation to the content: from simple to complex skills and concepts.

Focusing only on transmission of the content of mathematics in this way can lead to teaching that emphasizes the abstractions, related procedures, and mathematical concepts without considering the learners' progressive cognitive development. In a framework like this, learning is understood to move along a line. Each lesson, each day, is geared to a different objective, a different "it." At the end of the lesson, all children are expected to understand the same "it," in the same way. They are assumed to move along the same learning path; if there are individual differences, it is just that some children move along the path more slowly—hence, some get classified as "slow" or "below-grade-level" learners needing more time or remediation.

To see the fallacies in this approach to instruction, let's imagine a far-fetched example—teaching a child to walk by breaking the activity into a set of skills and working on each until mastery is achieved. "First, the right foot . . . watch me now. Balance on the left and lift the right. Practice now, over and over, until you can do it without falling. Good job!" Heap on the praise for reinforcement and then check off that skill as mastered! "Now, the next skill . . . on to the left foot. Up, down, up, down . . . there! Mastery of that skill achieved, too. And now finally we are ready for Bloom's level of synthesis. Let's put the skills together and try walking across the room." Just as ridiculous is the idea of doing a hands-on activity with a group of toddlers and expecting that at the completion of it they should all "get it" in the same way at the same time.

Those who have spent time with young children know that neither approach is sufficient to engender learning precisely because learning *is* development. It is the child's desire and inquiry to stand and get across the room that encourages the first faltering steps. The interaction between maturation and the social surround—the fact that the other humans around him walk upright—also affects this development. Studies of a few children who spent many years deprived of human contact found that they may not walk totally upright. Instead, they developed a gait characteristic of the interactions they witnessed or dictated by the limitations of their environment (Candland 1993).

When toddlers take their first faltering steps, we facilitate their development by celebrating their attempts and upping the ante when the first step is taken. We hold our hands out but most likely take a step backward. We make the surround as rich and enticing as we can to support development. We notice the phenomena that prompt inquiry and place these objects (or people) within range. We capitalize on the human instinct to reach beyond one's grasp, and we celebrate the developmental landmark accomplishments.

"It is a general insight," Noam Chomsky has written, "which merits more attention than it receives, that teaching should not be compared to filling a bottle with water, but rather to helping a flower to grow in its own way. As any good teacher knows, the methods of instruction and the range of material covered are matters of small importance as compared with the success in arousing the natural curiosity of the students and stimulating their interest in exploring on their own. . . . We should not be speaking *to* [learners], but *with* [them]. That is second nature to any good teacher" (Chomsky 1988).

One could object to the example of learning to walk, saying it's far-fetched and related more to physical than mathematical development. But doesn't it have implications for learning in general? A child comes into this world not as a blank slate but with reflexes that soon become differentiated and integrated into schemes to act on the environment. We are born with the mathematical ability to recognize small amounts such as one, two, or three, to determine magnitude, and to imagine number spatially (Dehaene 1997). We come to "know" our surround further through exploration, interpretation, and construction—organizing and generating ideas into efficient neural networks of cognitive structures.

It is human to inquire. It doesn't take a biologist or cognitive scientist to confirm what the philosopher Philo stated long ago: Learning is by nature curiosity. However, inquiry is also social. It takes place within a cultural community of discourse and reflection and uses tools and forms of representation and argumentation characteristic of the specific discipline. As scientists experiment and collect data, they use cultural tools, ideas, models, and systems of measurement previously constructed by their community. They share data in juried publications. Various scientists often work on small pieces of large communal puzzles. The same is true of mathematicians. At a certain point enough of the pieces have been placed and an overall structure begins to emerge. Sometimes this new structure is even a paradigm shift—the tipping point has been reached, and past ideas need to be reorganized.

As teachers, do we see our role as initiating learners into mathematical communities, speaking and inquiring *with* young mathematicians at work? Do we open the doors of our community, respecting each learner as an apprentice fellow inquirer? Or do we speak *to* them, trying to transmit a set of skills and concepts arranged on a continuum based on an analysis of the discipline by previous mathematicians? Are we teaching the history of mathematics rather than mathematics? Or are we teaching mathematics as the alive, creative, generative activity it is? By inviting young children to solve problems in their own ways, we are initiating them into the community of mathematicians who engage in structuring and modeling their "lived worlds" mathematically.

Teaching to support the cognitive development of apprentice mathematicians is surprising and exhilarating. It is also complex and demanding, because it is not aimless and random. We have a critical, very important role to play as teachers. We walk the edge between the structures of mathematics and the development of the child. This means we have to understand thoroughly the *development of the mathematics* by considering the progression of strategies, the big ideas involved, and the emergent models that potentially can become powerful forms of representation with which to think.

Math teachers do not walk into the classroom wondering what to do, waiting for learners to inquire. They plan lessons and know what they expect their students to do. The investigations they introduce allow for many entry points, and as learners respond, they acknowledge the differences in their thinking and strategies and adjust their questions and comments accordingly. While they honor divergence, development, and individual differences, they also identify landmarks along the way that grow out of their knowledge of mathematics and mathematical development. These landmarks help them plan, question, and decide what to do next.

THE ALGEBRA LANDSCAPE: DESCRIBING THE JOURNEY

In previous work we have delineated this developmental trajectory, this "landscape of learning," in three areas: (1) early number sense, addition, and subtraction (Fosnot and Dolk 2001a); (2) multiplication and division (Fosnot and Dolk 2001b); and (3) fractions, decimals, and percents (Fosnot and Dolk 2002). But what is the landscape of learning for algebra?

DENSE AND SPARSE STRUCTURES

Imagine the number 64. How do you think about it? As a counting number that is one more than 63, additively as $60 + 4$ (and other equivalent expressions such as $59 + 5$, $58 + 6$, $70 - 6$, etc.), or multiplicatively (using place value such as 6×10 plus 4)? Or perhaps as an even number, a double (2×32), a square number (the multiple of 8×8), or as a power of two (i.e., 2^6)? Maybe you think of it geometrically, as a $4 \times 4 \times 4$ cube or as b^2 in a 6, 8, 10 right triangle ($a^2 + b^2 = c^2$) or as the number of hexagons in a fullerene that has 76 faces or as the number of faces in two truncated icosahedra (each having 20 hexagons and 12 pentagons, as in a soccer ball). If you've read the first chapter, you may be thinking of it as a seven-factor number. You have many ways to think about it, and the more ways you have, the denser your network of relations.

When cognitive structures are dense, they consist of many interconnected pathways, many networks of relations. Sparse structures, on the other hand, have few. Dense cognitive structures are important because there are more relationships to exploit when solving problems. The mathematician Keith Devlin has written:

> What makes it possible for the mathematician to see what to do . . . is that she or he sees an underlying structure to the problem domain. When you can see that structure, it often is obvious what to do next. Mathematical knowledge is not a collection of isolated facts. Each branch is a connected whole, and there are links between many of the branches. Think of it as an undulating landscape, much of it heavily forested and shrouded in mist. Trying to find your way around by trial and error is unlikely to get you to

your destination. It helps to know as much of the overall topography as possible so you can find the best route. Those individuals who seem to just "know" how to solve math problems have simply spent enough time exploring the mathematical landscape to have developed a good sense of the terrain" (2003, 33–34).

Sometimes it is helpful to peel off layers of relationships to expose a sparser structure, as this may be more illuminating to the problem at hand. Having dense structures allows for various "peelings." Mathematics taught as an isolated set of skills and concepts rarely becomes part of an interconnected network of relations. That's because structuring requires cognitive reorganization on the part of the learner; the ideas usually cannot be transmitted by explanation alone.

DEVELOPING DENSE STRUCTURES

Structures are the result of organizing elements into a system of part/whole relations. The parts of the system can be described in relation to one another as well as to the whole. Throughout the history of mathematics, many structures have been developed as mathematicians have structured new relations and built on previous ideas.

One of the earliest structures was the set of counting numbers, $N = \{1,2,3,4...\}$. Emerging from the need to count items, starting with 1, each number is related to the one before it by +1, or the one after it by −1. The counting numbers are ordered, and each number is either greater than or less than other numbers. So the set N has an order relation, which is symbolized $(N,<)$. In time, the drive for increased utility led to the inclusion of 0, both to facilitate numeration, as a placeholder in a place value system, and as an amount such as $3 - 3$ (Guedj 1997).

Operations and properties were soon added. For example, consider the operation of addition, $(N,<,+)$. Here, all the true addition sentences join the structure of the counting numbers: $1 + 1 = 2$, $3 + 5 = 8$, and so forth. Two parts add up to a whole, and thus when one part is removed the other remains.

This new structure also includes the algebraic relationships of the commutative and associative properties and strategies like compensation. For example, the use of compensation to find the sum of $63 + 98$ by using $61 + 100 = 161$ is a generalizable algebraic strategy that arises through structuring $(N,<,+)$ and the understanding that $(a + b) + c = a + (b + c)$.

It is also possible to encode forms of multiplication in the additive structure $(N,<,+)$. These are forms of multiplication by a given natural number and arise from thinking about multiplication as repeated addition. Here one thinks of $3 \times 4 = 3 + 3 + 3 + 3$ and $4 \times 3 = 4 + 4 + 4$. But suppose one only views multiplication as an additive structure. How is one able to understand $^3/_5 \times 15$ or understand a justification for the commutative law, that $a \times b = b \times a$? A multiplicative structure is more encompassing than an additive structure. Proportional reasoning is added. If you want to multiply by eight, you can double three times in succession, or if you want to multiply $3^{1}/_{2} \times 14$, you can double $3^{1}/_{2}$ and halve 14 and realize the product is the same as 7×7. Structuring the multiplicative aspects of the numbers in this way, one is employing the associative law of multiplication.

Attempts were made in the new math movement of the sixties to align instruction with vertical progressions like the above. Materials such as base blocks and Cuisinaire® rods were designed and used in an attempt to expose learners to the structures of mathematics (or to help them discover them). Learners were introduced to counting first, and then the rods were used to represent the cardinal sets. Addition was represented as the union of (nonintersecting) sets, and the increasing order in $(N,<)$ was explored by building staircases with the materials. Multiplication was introduced by arranging the materials into rectangular arrays to make the part/whole relations of the distributive property explicit. This approach failed for many reasons, and led Freudenthal (1991) to comment that mathematics should be thought of as a human activity of "mathematizing"—not as a discipline of structures to be transmitted, discovered, or even constructed—but as schematizing, structuring, and modeling the world mathematically.

To understand the development of structuring in learners it is helpful to consider two dimensions—a vertical one and a horizontal one (Treffers 1987). The vertical dimension progresses by mathematizing mathematics—that is, adding operations (for example, from counting to addition to multiplication to exponentiation, and so on, as described earlier). It depicts the development of more or less sophisticated mathematical processing. In contrast, the horizontal dimension makes a problem field accessible to mathematical treatment. It allows life to be treated mathematically; we stay within the realm of realistic contexts, building connections across problems. For example, one might envision a rectangular grid of rows and columns, the volume of a box, or the possible combinations made with three shirts and two pairs of trousers as situations that can be mathematized with the operation of multiplication. Structures become denser as growth occurs in both dimensions (see Figure 2.1).

FIGURE 2.1
*The Growth of
Dense Structures*

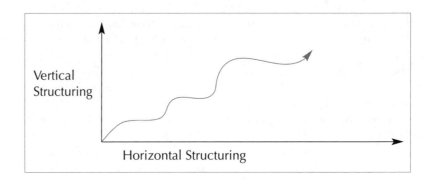

Another way to imagine the development of dense structuring is as a journey along a landscape. The vertical dimension is represented as a direct, linear, forward path toward the horizon, the horizontal as the many pathways and landmarks to the right and left. Traversing many paths over the landscape—traveling both horizontally and vertically—will result in a richer understanding of the terrain than if one takes only a direct path.

LANDMARKS ON THE LANDSCAPE

As we traverse the landscape of algebra development with young learners, there are many strategies, big ideas, and ways of modeling to notice and support. Signposts on the journey, these landmarks help you situate your learners developmentally. A graphic depiction of the algebra landscape described in this book is provided in Figure 2.2, although the landmarks on it will likely have meaning only once the whole book is read. Examine it only briefly at this point, continue reading, and return to it at the end.

Strategies

In Figure 2.2, strategies are depicted as rectangles. These are the forms of organizing, the schemes you will see as learners structure their activities. For example, initially learners may just use procedural arithmetic to determine if the statement $8 + 2 + 5 = 7 + 8$ is true. They add up both sides of the equation and if the answers are the same ($15 = 15$), they deem the statement true. Later in their development they prove the same statement employing associativity and commutativity: $2 + 5 = 7$, and since $8 + 7 = 7 + 8$ the statement is true. Still later they may disregard the eights, since they are on both sides of the equation, saying the statement is true because $2 + 5 = 7$. This strategy is later extended into an undoing strategy and employed to solve for unknowns $3 + 8 + 2 + 5 = n + 8 + 7$. The expressions $8 + 2 + 5$ and $8 + 7$ are removed by the addition of negative integers, leaving $3 = n$. The developmental progression of strategies, or "progressive schematization" as Treffers (1987) calls it, is an important inherent characteristic of learning.

Big Ideas

Underlying the developmental progression of strategies is the construction of some essential big ideas. These are depicted as ovals in Figure 2.2. Big ideas are the "central, organizing ideas of mathematics—principles that define mathematical order" (Shifter and Fosnot 1993, 35). As such, they are deeply connected to the structures of mathematics. They are, however, also characteristic of shifts in learners' reasoning—in logic—in the mathematical relationships they set up. As such, they are connected to part/whole relationships—to the structure of thought in general (Piaget 1977). That's precisely why they are connected to the structures of mathematics. As mathematical ideas developed

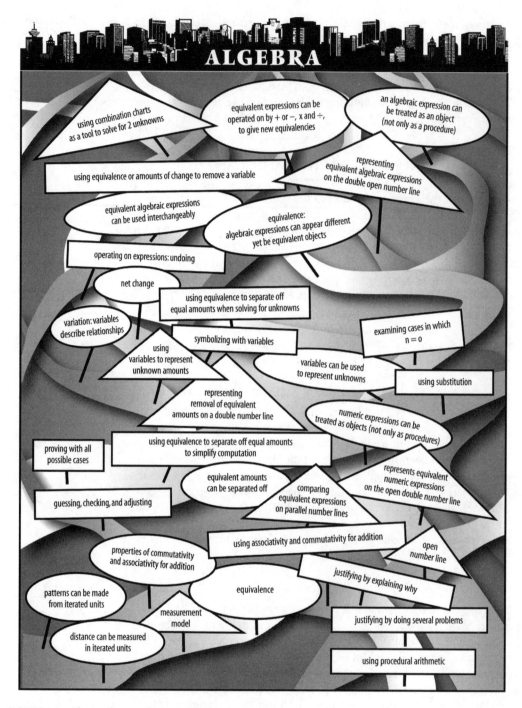

ALGEBRA

using combination charts as a tool to solve for 2 unknowns

equivalent expressions can be operated on by + or −, x and ÷, to give new equivalencies

an algebraic expression can be treated as an object (not only as a procedure)

representing equivalent algebraic expressions on the double open number line

using equivalence or amounts of change to remove a variable

equivalent algebraic expressions can be used interchangeably

equivalence: algebraic expressions can appear different yet be equivalent objects

operating on expressions: undoing

net change

using equivalence to separate off equal amounts when solving for unknowns

variation: variables describe relationships

symbolizing with variables

examining cases in which n = 0

using variables to represent unknown amounts

variables can be used to represent unknowns

using substitution

representing removal of equivalent amounts on a double number line

numeric expressions can be treated as objects (not only as procedures)

proving with all possible cases

using equivalence to separate off equal amounts to simplify computation

represents equivalent numeric expressions on the open double number line

equivalent amounts can be separated off

comparing equivalent expressions on parallel number lines

guessing, checking, and adjusting

using associativity and commutativity for addition

open number line

properties of commutativity and associativity for addition

patterns can be made from iterated units

equivalence

justifying by explaining why

measurement model

justifying by doing several problems

distance can be measured in iterated units

using procedural arithmetic

FIGURE 2.2   *The Landscape of Learning for Algebra*

through the centuries and across cultures, the advances were often character-ized by paradigmatic shifts in reasoning. These ideas are considered "big" because they are critical to mathematics and because they are big leaps in the development of children's reasoning.

For example, research has shown that many students comfortable with $x + 3 = 8$ do not understand how to interpret the expression $x + 3$ as an object by itself (National Research Council 2001). The confusion may be that they don't know how to combine the x and the 3 or that they believe that every time they add two things the result has to equal something else. Middle school mathematics teachers share stories like this repeatedly: "I gave my students a quiz and they were asked to factor $3x + 36$. Many students wrote $x = 12$." The notion that the expression $3x + 36$ can be considered an object in and of itself has not been formed—it is as if whenever the student sees an x they think they must find its value. Yet often, in algebra, mathematicians want to work with expressions (like $x + 3$) in a meaningful way. In other words, they want to treat the expression *as an object*. Indeed, it is common to find middle school mathematics texts (and teachers) diligently drilling students on the distinction between an *equation* and an *expression*, evidence that teachers of algebra have long recognized the critical importance of this big idea.

Unfortunately, such a big idea cannot be reduced to applying syntactical rules—it is a huge shift in thinking. Young students find it natural to solve for unknowns in equations. Even in first grade, students can think about ques-tions like __ + 3 = 8 and determine how to fill in the blank. Later, teachers may write $x + 3 = 8$ and tell the students "to find x," and again most students find this easy to think about. They see the question as what number is 8 three more than, or they count backward from 8 (removing 3) and then write $x = 5$. Here students see the use of the variable to represent an unknown value, and they understand $x + 3$ as an expression *to describe a procedure*.

An added complexity is that in teaching *variation* the procedure may remain the focus of attention and actually hinder the development needed for later algebra. Teachers prompting middle school students to think about why $2(n - 1)$ and $2n - 2$ were equivalent expressions asked, "How would you prove that $2(n - 1) = 2n - 2$ is true for all numbers?" (Boaler and Humphreys 2005, 66). The difference may seem subtle, but it is criti-cal. Does $2(n - 1)$ only have meaning when you can consider evaluating it for all numbers, or is it an object one can manipulate in and of itself?

Let's take another example, this time solving for two unknowns when presented with simultaneous equations: $x + y = 53$ and $x - y = 23$. If one thinks about $x - y$ and $x + y$ as two objects (the former being equivalent to 23, the latter to 53), then adding equivalent objects, one to each side of the equation also maintains the equivalence. So let's add $x - y$ to one side and 23 to the other: $(x + y) + (x - y) = 53 + 23$. This strategy simplifies the whole thing to $2x = 76$. Isn't that nice?

One could also envision the expressions as objects on a number line. Starting on a spot marked x and jumping a length of y, results in landing at 53. On the other hand jumping back a distance of y, results in landing at

23. This means *x* is at the midpoint between 23 and 53. The midpoint can be derived by finding the average: (23 + 53)/2 (see Figure 2.3).

Another big idea is related to understanding variation: *variables describe relationships—and are not merely unknown quantities.* In order to make sense of algebraic equations such as *x* + 3 = *y* − 2, students need to construct the idea of variation: An indeterminate amount can be related to another indeterminate amount, and this is meaningful even if the amounts are not known.

A third big idea is the understanding of equivalence: *Algebraic expressions can appear different yet be equivalent objects.* These ideas will be revisited and further clarified and many more big ideas will be described throughout this book.

Models As Tools for Thought

When we construct an idea, we want to communicate it. Through time and across cultures humans have developed language as a way to do so. Initially, language represents ideas and actions; it is a *representation of thought.* Language also serves as a *tool for thought.*

Numerals were developed to signify the meaning of counting. Operational symbols like × and ÷ were constructed to represent the actions of combining and portioning equivalent-size groups. Variables arose to signify variation, unknowns, and relations. Ratio tables, combination charts, arrays, and open number lines were developed as ways to represent relations graphically. While these symbols and models were initially developed to *represent* mathematical ideas, over time they become *tools*—mental images to facilitate thinking.

Professional mathematicians do a great deal of abstract mathematics by using models as tools for thinking. Consider the well-known Pythagorean theorem. Many adults will likely say, "Yeah, isn't that a-squared plus b-squared equals c-squared?" But do they realize this theorem is critical to finding distance in mathematics or understand why it should be true? Not as likely. In some geometry class they might have seen a depiction of a right triangle with squares added to each of its three sides (see Figure 2.4a). But likely this mysterious figure wasn't very helpful. It might visually clarify the question, "How are a-squared, b-squared, and c-squared related?" but there is no evident reason why these three squares should be related, nor does the diagram provide much motivation to care about them.

FIGURE 2.3
Midpoint Problem on a
Number Line Showing
x + *y* = 53
x − *y* = 23

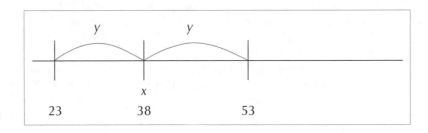

In contrast, consider the two large squares in Figure 2.4b. They are dissected in interesting ways, and we see four right triangles in each large square, each triangle with side lengths a, b, c. Can you find squares with areas of a^2, b^2, and c^2? Look at the part-whole relations between these squares in the two larger squares, both of which are the same size. Can you see now why the Pythagorean theorem is true? To be sure, the squares in Figure 2.4b have been carefully dissected to facilitate this understanding.

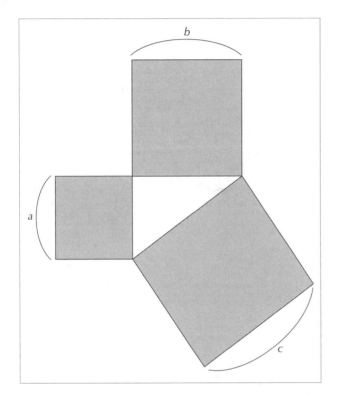

FIGURE 2.4a
*Representing the
Statement of the
Pythagorean Theorem*

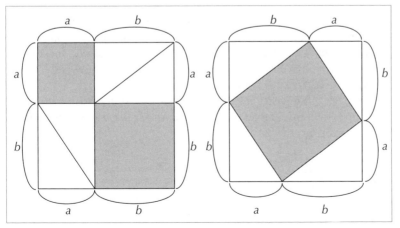

FIGURE 2.4b
*Representing a Proof
of the Pythagorean
Theorem*

This is what mathematicians do; they find ways to represent relationships to illuminate important ideas.

Of course, this example is likely too difficult for young children. You may have even skipped this section yourself! Therefore, this model is not included on the portion of the landscape of learning shown on page 30. You will, however, see the double number line included as a landmark in several places. It is a powerful tool for thinking algebraically—specifically for representing variation and expressions as objects that are related, solving for unknowns, exploring common multiples and factors, and thinking proportionally. For example, imagine three jumps of the same size along a track, compared to four smaller jumps of equal length that end at the same point (see Figure 2.5). We might represent this relation as $3x = 4y$. What can you tell about the lengths of the jumps? Can you predict how many more jumps of each are needed to meet again? Can you tell how many jumps of x would be needed to match the length of 2 jumps of y? Or just one? What if you learned in addition that $3x + 4y = 24$? Does this information force specific values for x and y?

The mental images learners form and then manipulate allow them to work meaningfully with the actions most people associate with the word *algebra*, such as symbolizing, symbolic reasoning, solving for unknowns, and so forth. Imagine a young child who is adding a pile of six beans and a pile of eight beans but miscounts and finds thirteen. If you tell the child, "That can't be right because six and eight are even and thirteen is odd," the child won't have a clue what you are talking about unless he has constructed the mental objects labeled as even and odd. Sadly, much of current algebra teaching fits this pattern: little to no structuring, little time to construct the mental objects, but lots of talk about structures and relations. This is why we have developed the landscape.

WALKING THE EDGE

A good teacher walks the edge between the structure of mathematics and the development of the child by considering the progression of strategies, the big ideas involved, and the emergent models. Ultimately what matters is the mathematical activity of the learner—how the learner mathematizes and structures the situations the teacher offers. But learning—development—is complex. Strategies, big ideas, and models are all involved, and they all need to be developed as they affect one another. And vertical and horizontal explorations are both critical.

Strategies, big ideas, and models, however, are not static points in a landscape—objectives to get everyone to the same place at the same time in

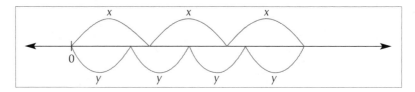

FIGURE 2.5
3x = 4y on an Open Number Line

the same way. They are dynamic movements on the part of the learner in a journey of mathematical development, and the role of teachers is to foster, support, and celebrate this development.

Several tools can facilitate this journey. One is to craft problem situations that beg to mathematized. Contexts like these, designed with potentially realizable strategies and/or built-in constraints to prior strategies, can be powerful inducers of development. Facilitating inquiry and dialogue, asking learners to read and comment on one another's mathematics, and holding math congresses to discuss key strategies and ideas also spark development. And last, short guided minilessons at the start of math workshop can introduce certain strategies for examination. In the next several chapters, stories from classrooms illustrate these possibilities.

SUMMING UP

Historically, learning has been seen as a linear vertical path depicting a progressive list of skills, concepts, and procedures—arrived at by dissecting the already formed structures of the discipline in an attempt to transmit (or enable learners to discover) them. Yet as Freudenthal points out, "Cognition does not start with concepts, but rather the other way around: concepts are the results of cognitive processes."

The development of dense cognitive structures over time allows learners a larger set of lenses when approaching problems. As learners develop, their structuring increases in two dimensions, both in the number of the operations being considered (vertical structuring) and in the interconnectedness of the relationships involving those operations (horizontal structuring). This goal makes teaching and learning very complex causing Freudenthal to comment, "How often haven't I been disappointed by mathematicians interested in education who narrowed mathematizing to its vertical component, as well as by educationalists turning to mathematics instruction who restricted it to the horizontal one." Freudenthal understood that narrow views of education, at either extreme, severely limit children's experiences. If we as teachers have a deep knowledge of the landscape—the big ideas, the strategies, and the models that characterize the journey—we can facilitate inquiry through contexts that support children's journey in both dimensions.

Einstein wrote, "It is nothing short of a miracle that modern methods of instruction have not yet entirely strangled the holy curiosity of inquiry" (Eves 1988, 31). But change is not easy! When the act of teaching is viewed as the "filling of a bottle with water" it is easy to quantify and measure the levels of the outcomes, easy to compare and label humans as fit and unfit, and easy to ensure the continuation of the status quo with its categories of "haves" and "have nots." In contrast, when teaching is characterized as "helping a flower to grow in its own way," due emphasis is placed on learning as development and as treating young children as developing mathematicians from the start. Equity and access are ensured; empowerment of all is the result.

3 | EARLY STRUCTURING OF THE NUMBER SYSTEM

Whenever a large sample of chaotic elements are taken in hand and marshaled in the order of their magnitude, an unsuspected and most beautiful form of regularity proves to have been latent all along.
—*Sir Francis Galton (1822–1911)*

Classification is not a matter of child experience as things do not come to the individual pigeonholed. The vital ties of affection, the connecting bonds of activity, hold together the variety of his personal experiences. The adult mind is so familiar with the notion of logically ordered facts that it does not recognize— it cannot realize—the amount of separating and reformulating which facts of direct experience have to undergo before they can appear as a study, or branch of learning.
—*John Dewey (1902)*

The first structuring of the number system (usually noticeable in young learners around the age of five or six) is the arrangement of the counting numbers as a sequence of amounts nested one inside the other. As children construct the idea that the result of their counting is an amount (in contrast to the name of the last object or just the last word in a counting song), they come to realize that the counting numbers increase in increments of +1 and therefore nest inside each other almost like Russian nesting dolls. Researchers call this idea *hierarchical inclusion* (Kamii 1985; Fosnot and Dolk 2001a). Another important big idea about number constructed early on is *one-to-one correspondence*—that if six children each need a partner, six more children are needed.

When addition contexts are posed, young children initially make sense of the situation by counting three times (once for each part, and then counting all). This provides the basis for early part/whole relations involving addition, namely the two parts (represented by the two addends) and the entire quantity (the sum). Around the age of six or seven, rather than counting three times, children begin to "count on." For example, to figure out 5 + 3, they count on from the 5, saying "six, seven, eight." Once they

understand the whole as comprising two disjoint[1] subsets, they can incorporate subtraction (as removal) into this mental construct and the full
breadth of part/whole relations emerges—that 5 + 3 = 8 and 8 − 5 = 3 are
two expressions describing the same mental image. They are equivalent
statements as they reflect the same part/whole relationships.

This is only the beginning of the additive structuring of number, and
some say this belongs in the domain of number, not algebra. However, this
early structuring also allows learners to structure the natural numbers in algebraic ways using additive properties, rather than simply recording the results
of addition and subtraction. For example, the subdivision of natural numbers
into two parts, the evens and the odds, is a structure that students recognize
early on as they work to automatize the basic addition facts. Doubling is an
important arithmetic action because it helps students construct new facts from
old (if 3 + 3 = 6 then 3 + 4 = 7), but it also helps students form the mental
image of the subdivision of the natural numbers into even and odd.

TEACHING AND LEARNING IN THE CLASSROOM

Madeline Chang is using the unit *Beads and Shoes, Making Twos* (Chang and
Fosnot 2007) to invite her young kindergarten and first-grade mathematicians to begin structuring the number system. To explore doubling, she
uses the context of walking hand-in-hand in two lines, which the children
are used to doing when they leave their classroom for other areas of the
school or take field trips. She reads the classic children's book *Madeline*, by
Ludwig Bemelmans, to develop the context and then asks her children to
draw different-size groups (lines) to produce doubles—class sizes that
allow everyone to have a partner. Enthusiastically, one of her five-year-olds,
Sofia, says a number aloud before they even leave the meeting area.

"I know one!" she exclaims. "When five kids have partners, it's ten,
because five plus five makes ten!"

Madeline smiles and writes, 5 + 5 = 10. "So 10 is a special number that
works. Let's find some more."

The children eagerly go off to work in pairs. Many begin by drawing
one line first and only then complete the second line. It is sometimes a
struggle for young children to draw two corresponding sets of the same
number of objects, and they might draw the second line of children longer
or shorter than the first line.

This is true for Josie, who suggests using plastic teddy bear manipulatives to her partner, Chloe. "Let's make lines with the teddy bears first, and

---

[1]*Disjoint* means the sets have no elements in common. If the sets are not disjoint, then they do
not model addition; to determine the total, counting all may be necessary rather than a counting on.

then we can draw the children." Josie then makes two lines of objects, one line longer than the other.

Madeline notices this and says, "I see that you are making beautiful lines, Josie. But I'm wondering—one of your lines looks longer than the other. How do you know that everyone in this line has a partner in the other line?"

Reiterating the context helps children realize the meaning of what they are doing and may create conditions for them to develop their own solutions. It can help them develop an understanding of one-to-one correspondence and doubling, landmarks that are precursors to being able to structure the natural numbers into evens and odds.

Josie looks puzzled but then figures out a solution. "I know! We have to hold hands with our partners. I can draw them holding hands!" She draws two rows of teddy bears and represents their holding hands by drawing a line from each bear in one row to the corresponding bear in the other row (see Figure 3.1). "We need one more teddy bear here."

Madeline asks, "How many do you have in each row now?"

Josie counts: "One, two, three, four, five, six, seven. I have seven!"

"And how many in this line?"

Madeline's question is powerful: She doesn't assume that Josie understands one-to-one correspondence. Josie's answer will provide important information, proving in fact that Josie has constructed that idea.

"That's seven, too! See, they're holding hands now."

Madeline celebrates this early landmark and then challenges the girls by introducing the addition. "I see. That was a great idea! So shall we record this one as 7 + 7? How many children is that?"

Instead of counting on from seven, Josie begins all over at one, counting up to fourteen. *She has not yet constructed an understanding of part/whole relations.* "How do you write that?" she asks as she finishes.

"Like this—14." Madeline shows Josie how to make the numerals (*how a number is represented is social knowledge—a label or name—and does not need to be constructed in the same way as a mathematical idea*) and then celebrates the finding of another even number. "Wow! You found another special double number."

After the children have worked for an appropriate length of time, Madeline asks them to prepare for a math congress by looking back over their work and making a list of all the doubles they found. She knows this will generate a large assortment of even numbers, and she plans to place these numbers on a long strip of paper (reminiscent of an open number line) and discuss any patterns the children notice. She hopes they will begin to notice patterns and suggest new doubles for the list. *The line of even numbers is not only a representation of their work but also a tool for thinking.* She also places a standard number line with the counting numbers 1–100 nearby, hoping children will notice that the standard number line has numbers going up by ones and the one the class is making skips every other number. *They may also notice that all the numbers on their line end in 0,*

2, 4, 6, *or* 8, *whereas the numbers in between—those not listed—end in* 1, 3, 5, 7, *and* 9.

"So it seems we found a lot of numbers. Let's start posting them on this strip, like a number line." Madeline points to the strip of paper she will be using and the standard number line displayed nearby. "I wonder which of these is the smallest. Where shall we begin? Mathew?"

"I did three plus three. It's six."

Madeline pushes three red beads over on the top row of the class arithmetic rack[2] and three red beads over on the bottom row. "Do we agree that

---

[2]For information on arithmetic racks see Fosnot and Dolk, *Young Mathematicians at Work, Constructing Number Sense, Addition, and Subtraction*, pp. 105–12, or go to www.mathrack.com.

FIGURE 3.1
Holding Hands: Josie and Chloe's Work

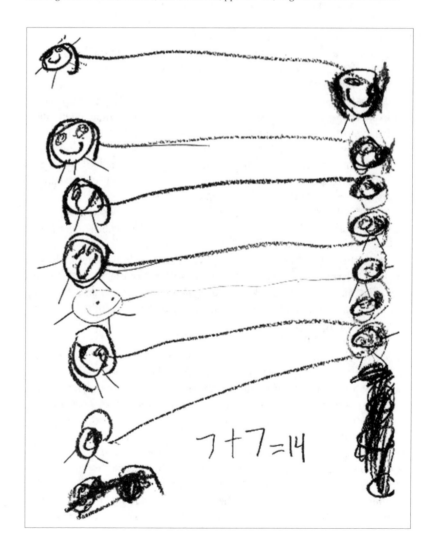

three plus three is six?" Some children count and then nod; others nod immediately. "Okay. Let's post your record, then."

"I don't think it's the littlest though," Leah pipes up. "I think we could do two and two. That's four."

"Shall we add that one?" *Madeline does not acknowledge the answer as correct but turns to the community for consensus. Since the numbers are small, the children can easily imagine them.* Again, many children nod.

It is common that students will not see 1 + 1, *or* 0 + 0 *as possibilities in this context because they don't think of one child (or 0) as a "line." For now Madeline does not push to add these to the list.*

"Daniel, you tried a different number. What did you find out?"

"I found out that four plus four equals eight."

"How shall I make that on the arithmetic rack?" *Madeline presses the children to envision arrangements of the amounts.*

"One more on the top and one more on the bottom," Daniel says with conviction.

Madeline restates what they have so far. "So two plus two was four, three plus three was six, and now four plus four is eight."

"Two more! It's two more every time!" Sofia exclaims. Other murmurs of surprise and delight are heard as well.

Madeline smiles and encourages the children to continue examining this relationship. "Isn't that interesting? Two more."

Sadie, another five-year-old, shows her drawing of fourteen children in two lines (see Figure 3.2), "I counted by twos. See . . . two, four, six, eight, ten, twelve, fourteen. It's seven and seven in a line. If it's eight kids, it would just be two more."

"Hey, look!" Mathew has jumped up in excitement and is pointing to the standard number line. "That one over there is going up like this—1, 2, 3, 4, 5—and ours is 2, 4, 6, 8. It's just skipping numbers!"

Madeline is delighted. "What an interesting thing to notice Mathew. Do you think then that we could add more double numbers to our list?"

"Yeah, 'cause it's like Sadie said—count by twos."

WHAT IS REVEALED

Madeline has succeeded in encouraging her five- and six-year-olds to begin structuring the number system into evens and odds, but their work so far is the very beginning of the journey. The children initially construct the one-to-one correspondence inherent in doubling contexts (represented by "holding hands"), but they move toward thinking about this additively. They are considering a "double number" as a number added to itself. Next, some children see the sequence as growing in a +2 fashion, when the increase by two represents a new pair of children holding hands. One of Madeline's goals is to help their structures become denser, so she needs to move both horizontally and vertically (as described in Chapter 2). She

wants to provide them with many more "horizontal" experiences exploring
even/odd relationships, and she also wants to encourage vertical moves.

For example, one of the big ideas related to even numbers is the multiplicative aspect of the structuring, where $2n$ means two *times* a natural
number (including 0). With this, odd numbers are represented in relation
to the evens as $2n + 1$, thereby being understood in terms of part/whole
relations. The commutative property for multiplication then explains why
Sadie's point holds: groups of twos, n times, equals 2 times any natural
number ($2n$). But is multiplicative structuring developmentally appropriate? What can be expected of these kindergartners and first graders? How
far along the landscape can they traverse when structuring is supported?

FIGURE 3.2
*Sadie's Work:
Counting by Twos*

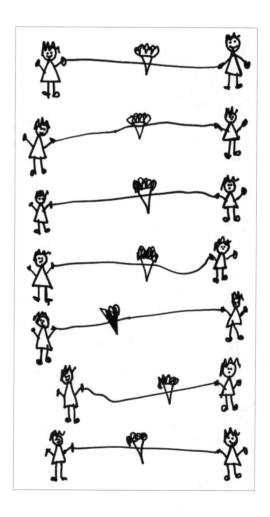

At first Madeline traverses the landscape horizontally. Over the next several days she makes a habit of beginning math workshop with brief minilessons—strings of related problems using doubles and near doubles. For example, she shows one problem at a time on the class arithmetic rack and asks for thumbs up when children have an answer. She explores alternative strategies but encourages the children to make use of the relations in the string.

Here's one string of related problems she presents:

Three on the top, three on the bottom
Five on the top, five on the bottom
Five on the top, six on the bottom
Eight on the top, eight on the bottom
Seven on the top, eight on the bottom
Six on the top, six on the bottom
Seven on the top, six on the bottom

The first two problems in the string are chosen with the expectation that they will be easy for the children. The second problem can be used to solve the third. The fourth problem is more challenging and many children resort to counting by ones, but Madeline encourages them to think about whether any of the previous problems can be helpful (the first two— $(5 + 5) + (3 + 3)$—for example; the color of the beads on the arithmetic rack—twenty beads arranged in two rows, each with five reds and five whites—makes these doubles stand out). The string continues with more doubles and near doubles, and children are encouraged to use a known double to solve an unknown near double.

Simultaneously, Madeline begins a new investigation related to the eggs the class has been incubating as part of their science work. The children have noticed that egg cartons come in various sizes (6, 12, and 24), all "doubles." Capitalizing on this, Madeline asks the students to think about why that might be and to design containers for larger numbers of eggs. How many eggs would a carton with two rows of 10 hold? How about one with two rows of 13? Why wouldn't there be a box for 7 eggs? Children make posters of their designs, display them around the room, and have a "gallery walk" so everyone gets a chance to see the various containers. (A few samples of work are shown in Figures 3.3a and b.)

After the gallery walk Madeline convenes a math congress. Using a pocket hundred chart, the children record the numbers by placing colored transparent inserts in the chart to highlight the sizes of all the containers they have made. Although not all the even numbers are highlighted, enough are for children to begin to notice some patterns and to make predictions of other numbers that might be doubles.

"Hey, all the numbers have 0, 2, 4, 6, and 8 in them," remarks Sadie.

Mathew proudly asserts his conjecture. "I think the numbers will always go down (*referring to the columns on the chart*). See, 2, 12, 22—it goes down. It skips 3, 13, 23. And over there it is 4, 14, and 24. I bet 34 is a double, too. Let's check."

"Yeah, it's gonna skip like that," Sadie nods in agreement.

EMERGING MULTIPLICATIVE STRUCTURING

As the children explore even and odd numbers, the variety of representations and contexts (number lines, pairs of children walking in line, egg cartons, a hundred chart) helps them notice different relationships. The

FIGURE 3.3a
Designing Egg Cartons

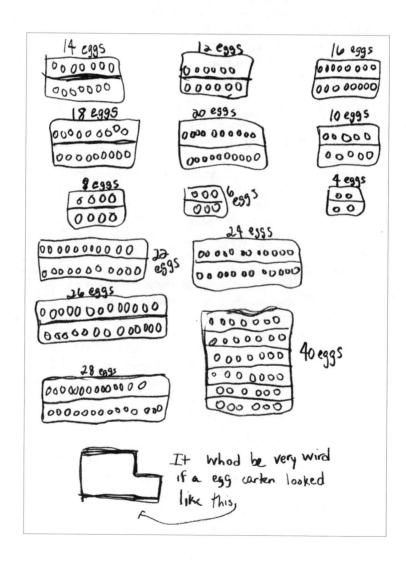

number line calls attention to the +2 increments, for example, and the hundred chart prompts a discussion of the 0, 2, 4, 6, and 8 digits. Their structures are becoming denser as Madeline provides horizontal structuring opportunities.

When Madeline realizes that she is beginning to notice many examples of doubles in her daily life, she asks the children whether they are too, challenging them to begin to mathematize their world—and introducing another investigation. She explains that last night she looked in her closet and realized—shoes! Shoes are perfect for thinking about doubles! Shoes come in pairs—and a pair is a double of one. How many shoes in five pairs,

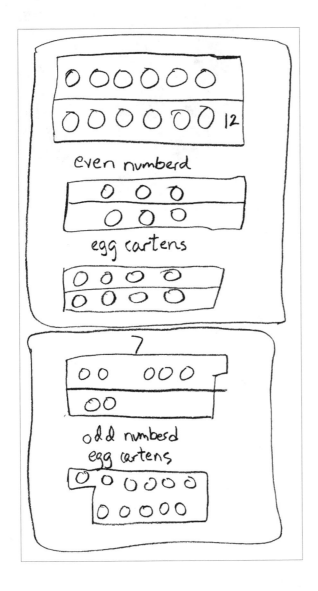

FIGURE 3.3b
*Designing Egg
Cartons (continued)*

twelve pairs, or fifteen pairs? She suggests they think of a group of people and the number of shoes that would be. (Some of the children's work is shown in Figures 3.4a and b.)

FIGURE 3.4a
Shoes

FIGURE 3.4b
Shoes (continued)

After the children have worked on their own for an appropriate length of time, Madeline convenes a math congress and places their combined data on a t-chart (see Figure 3.5).

"So now we have a chart with all our findings. After mathematicians collect all their findings, they often put them on a chart like this. Then they step back and think, 'Is there anything interesting here?' So mathematicians, turn to the person next to you and talk about anything interesting you notice." After several moments, Madeline resumes the whole-group conversation. "Ethan, what did you and Josie talk about?"

"All the numbers of shoes are over here." Ethan points to the open number line constructed when they explored children walking in line with partners. *He has noticed that all the answers are even numbers. For a young mathematician, this is an important observation.*

"Who else noticed that?" Several hands go up. "Why would that be happening, I wonder?"

Sofia offers her earlier insight about increments of +2. "Because you go 2, 4, 6, like that. You skip-count."

"Two shoes in every pair," says CJ. *He is beginning to unitize groups of two as one, given the pairing context.*

"You can't get a different number—you would have a shoe missing," Josie offers. "If you go 2, 4, 5, the 5 would be for only one shoe. A shoe is missing."

"So I see that it is six pairs of shoes, and that is 2, 4, 6, 8, 10, 12. But how did it end up being two sixes, like our double numbers?" *Madeline challenges her students to consider the equivalence of 6 twos to 2 sixes—to examine a case of the commutative property for multiplication.* "Will this always work? If we have five pairs of shoes, will it be two fives?" *Madeline purposely uses the language of unitizing—making the group a unit to be counted.*

Several children say, "No, not always—just sometimes."

Daniel is bursting with excitement. "It will, it will! Always! And I know why!" *Daniel sees the beauty of this emerging multiplicative structure.* He produces his picture (see Figure 3.4b). "I did six pairs of shoes. But I thought of it as six right shoes and six left shoes." Most of the children are

Number of Pairs of Shoes	Number of Shoes
1	2
2	4
3	6
4	8
5	10
6	12
7	14
8	16
9	18
10	20

FIGURE 3.5
T-Chart Illustrating the Number of Shoes for Groups of People

puzzled by this, but Daniel explains that if the shoes are arranged in pairs in a line, they can be seen as two lines, one line of right shoes and one line of left shoes. Eventually most of his classmates begin to grasp the multiplicative relations he has noticed.

"So numbers that go in twos also can be thought of as doubles? And the other numbers, the ones that don't work, don't go in twos? There is an extra one?" *Madeline is encouraging her students to examine evens in relation to odds—as part/whole relations in the set of counting numbers.* "Actually, mathematicians have a name for these numbers. The double numbers they call even numbers. The ones that don't go in twos—the ones that have an extra—they call odd numbers." *She introduces the terminology of odds and evens, but only after children have constructed the ideas for themselves.*

Josie finds this terminology quite logical. "Yeah, it would be odd. If you have five shoes, you know something is wrong. A shoe is missing!"

Michael has been looking at the hundred chart on which the egg carton data is recorded. "Look. I just noticed something. On the hundred chart—it goes odd, even, odd, even, like that. It's a pattern."

WHAT IS REVEALED

These five- and six-year-olds are forming a variety of representations of odd and even numbers. They are examining the natural numbers in relation to one another and noting the alternating pattern. They realize that even numbers can be thought of in two ways—as doubles ($2n$) and as n sets of two—and that odd numbers have one single more ($2n + 1$). Although most of the children are still most likely structuring these relations additively, multiplicative ideas are slowly beginning to emerge. The two-for-one unit (two shoes in a pair) is seen in relation to twice the number of pairs (see Figure 3.6). The arraylike arrangements (rows and columns) of lines of children holding hands, egg cartons, and pairs of shoes lined up are helping children visualize these relations.

FIGURE 3.6
*Multiplicative
Structuring:
$5 \times 2 = 2 \times 5$*

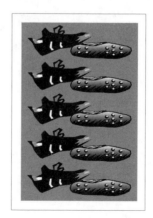

Structuring the natural numbers into evens and odds, however, is not the end of the journey; it is only a landmark idea along the way. The real goal is the *development of the structuring*—an infinite journey. The journey of these young mathematicians is thus far mostly horizontal. Even and odd numbers have been explored in a variety of situations (lines, cartons, shoes) and with a diverse set of representations (number lines, hundred charts, arrays, ratio tables). This horizontal structuring has brought up several patterns and relationships. A small vertical step has been taken in the discussion of the unitizing of groups of twos and the relationship of five pairs to two fives ($5 \times 2 = 2 \times 5$).

BACK TO THE CLASSROOM

To continue this vertical movement Madeline reads the picture book *Grandma's Necklaces* (Fosnot 2007b). In the story Grandma has discovered some very special numbers that help her make three different necklace patterns using blue beads and green beads. Her granddaughter wonders how her grandma knows what numbers of beads will work for each necklace. The mystery of the numbers is not solved in the story, so it is a great introduction to a math investigation.

The first necklace is made by alternating blue and green beads. Therefore it can be made in various sizes with *n* sets of two—one of each color. The repeating unit is one blue and one green bead, and as long as the numbers of blue and green beads are equal, any reasonable number of beads (10, 12, 14, and so on) works. However, if there is an odd number of beads, two beads of the same color will then appear next to each other when the string is tied.

The second necklace is a bit more challenging; the repeating unit is five blue beads and five green beads, or ten beads. Although all possible numbers of beads for this type of necklace must also be even, the added constraint is that they must also be multiples of ten. For example, 20 beads work (two sets of five blue and five green) but 25 beads do not (because there would be ten beads of one color together when tied).

The third necklace is made with a repeating pattern of three blue beads and three green beads; therefore, it can be made only with even numbers of groups of three, which is the same as saying that only multiples of six will work.

Madeline reads the book and then challenges her young mathematicians to figure out Grandma's special numbers. She doesn't expect all the children to solve all the necklace problems, so she uses an open-ended format that will support each child as far as he or she is able to go. The intent is to give every child rich opportunities to continue structuring the number system.

Nate examines the numbers in the list that he and Sofia have developed for the first necklace, the one with the alternating blue and green beads. "I think there is a pattern. It's gonna be every other one again. Like before."

Sofia agrees, "Yeah. Like when we walked in line."

"Say more about that. What do you mean? Are you noticing something about how the necklaces are related to walking hand-in-hand in line? *Madeline is requesting elaboration to make the relationships more explicit.*

"Yeah. It's like the blue and green beads are holding hands. They go together, blue and green, blue and green," Sofia explains.

Nate nods. "It's like the shoes, too. The blue and green are like a pair of shoes. They go together, like Sofia said."

Madeline pushes them to generalize. "So are you saying that any number of green beads will work as long as the greens have the same number of blues? And when I add these numbers of greens and blues together, no matter what the number, I get an even number?"

Sofia is convinced. "Yes. You can make the necklace in lots of sizes, and the sizes will all be even numbers. Odd numbers can't work, like the shoes. You need a pair. If you have one extra, it's a problem. The greens have to be the same number as the blues."

The second and third necklace patterns with alternating groups of five (necklace two) and alternating groups of three (necklace three) are a bigger challenge, pushing children to consider not only even numbers but even numbers of groups. *This is a difficult idea for young children because here the group is being considered as a unit. It is just this challenge, however, that can be instrumental in developing multiplicative structuring—a vertical movement.*

Initially many children use a trial-and-adjustment strategy. They select numbers to try at random and when one doesn't work, they add or remove beads to try to make the pattern fit. A few children skip-count by fives or threes, make a list of those numbers, and are surprised when not all their numbers (multiples of five or three) work. Madeline encourages them to circle the numbers that do work, and they eventually notice that every other number on their list works (even numbers of groups).

Some children are intrigued by how some number combinations result in necklaces with a long section of beads of the same color and others don't. For example, for the second necklace pattern, ten green beads and fourteen blue beads create a section of nine blue beads in a row when tied into a circle. On the other hand, if one uses six green beads and ten blue beads, there will only be one green bead between two sets of five blue beads.

As the children identify numbers that work, they place colored transparent inserts over those numbers in the hundred chart. A few children notice that all the numbers that work for the second necklace pattern are in the tens column (10, 20, 30, 40, etc.).

Sofia announces a related connection. "Hey! Like walking in line again! Every time the green gets five more, the blue does, too."

Daniel expands on Sofia's idea. "Yeah, it's like the blue and green beads are holding hands. They go together, blue and green, blue and green, but now there are five in each line. Four groups of five worked, then six, then eight. The fives have to be even. Odd ones don't work."

"Do you mean the number of groups has to be even?"

"Yeah . . . because they go together—two colors, five greens and five blues—and that makes an even number. That makes ten!"

WHAT IS REVEALED

As Madeline progressed through this unit, her young learners developed along two dimensions: horizontal and vertical. Initially most of the activities were designed to support horizontal structuring. The horizontal dimension allows the real world to be treated mathematically. Madeline used lines, egg cartons, shoes, and bead necklaces as contexts to examine odd and even relationships. She used number lines, hundred charts, arrays, and t-charts to represent the data in various ways to help students notice a number of patterns.

The vertical dimension "mathematizes" the mathematics—that is, it moves students to a more sophisticated mathematical process. Madeline helped her students move from counting by ones to using a two-for-one relationship as they counted pairs. Initially they set up one-to-one correspondences by counting, but because of the contexts they quickly began to use additive structuring as they formed doubles ($n + n$) and near doubles ($n + n + 1$). Multiplicative structuring began to emerge when they realized that pairs of shoes could be thought of in an array—as columns of left and right feet ($2n$) and rows of pairs (n sets of twos). The bead necklace investigations continued to support and challenge children to operate with even and odd numbers by examining groups of numbers.

As progressions occurred along the two dimensions, students' structures became denser. They explored and developed a variety of relations. This density promotes generalizations, which gets to the heart of algebra. Here are some of the generalizations the students articulated as they explored the three necklace patterns:

- Any numbers that work for the second and third styles also work for the first.
- Only even numbers work because there are two colors and the beads need to "hold hands."
- Odd total numbers of beads don't work for any of Grandma's patterns.
- More numbers work for the first pattern than for any of the others.
- For a number to work for the first pattern, you have to land on it when you skip-count by twos.
- For a number to work for the second pattern, you have to land on it when you skip-count by tens.

The necklace investigation has generalizing potential for older students as well. Children in second and third grade will spend more time with the third pattern, noticing that even numbers are necessary but also noticing the role of six in these special numbers:

- For a number to work for the third pattern, you have to land on it when you skip-count by sixes.
- For a number to work for all three patterns (Grandma's very special numbers), you have to land on it when you skip-count by twos and tens and sixes.
- If you skip-count by thirties, you get all the really, really special numbers.

FURTHER STRUCTURING: EVENS, "THREEVENS," AND PLACE VALUE

At this point, students in Madeline's class have informally been adding even and odd numbers to produce even numbers. They have not explored the formal aspects of additive operations with even and odd numbers—that is, if you add two even numbers you get another even; if you add two odd numbers you get an even number; the sum of an even and an odd is odd. However, it is common for teachers to ask students to articulate reasons for why these statements are true. (See Carpenter et al. 2003, 116–18, for a discussion of how students might approach this using multiplicative structuring.)

To deepen students' understanding of the part/whole relations involved in the even/odd structure it often helps to explore a variety of ways to structure numbers and compare the results. For example, an extension of even-odd structuring appropriate for older children is to partition numbers that arise when you divide by numbers other than two. For example, you might consider "threeven" numbers, those you get when you skip-count by three. Children note that the numbers 1, 2, 4, 5, 7, 8 are not "threeven" numbers, while 3, 6, 9, 12, and so on are; and as they structure this set of numbers, properties quite different from those that arise with even and odd emerge. In the following vignette, fifth graders Sandra and Marcia are investigating "threevens."

"Hey, this isn't good," Sandra declares. "If you add threeven numbers you get another threeven number, but if you add two not-threeven numbers you can get a not-threeven number."

"Yeah," Marcia says with chagrin. "It's like odd plus odd equals even doesn't work anymore. Two plus two isn't threeven."

"This isn't good."

"But one plus two is threeven . . . and so is four plus five."

Sandra offers a tentative conjecture. "Maybe threevens next to each other—no, I mean the ones that aren't threevens next to each other. . . ."

"Let's write that. Non-threevens next to each other add up to a threeven."

Bill, their teacher, presses them to reflect on the part/whole relations. "That's interesting. When do you think non-threevens add to a threeven and when do they not? Could you generalize this idea?"

What is emerging in this inquiry, in contrast to the earlier explorations in Madeline's class, is an important idea involving divisibility. These fifth graders will come to realize that when a number is divided by 3, the remainder will have to be 0, 1, or 2, and that when threevens are divided by 3 the remainder is 0. Initially ignoring the operation of division opens the door to structuring the addition of threevens horizontally. In time, Marcia and Sandra will recognize that when you add a non-threeven with remainder 1 to a non-threeven with remainder 2, a threeven results. Ultimately they will structure numbers according to three subsets (not two, as with even and odd), laying a foundation for modular arithmetic.

The children in Madeline's class will make similar observations when they connect place value with multiples of numbers. They will notice that even numbers all lie in the same column of the hundred chart. As they go down the hundred chart they are adding tens, and if n is even, so is $n + 10$. Ultimately it is all bound up in an early form of the distributive property: If two numbers are divisible by two (or three, or five), then so is their sum. This can be expressed symbolically as $2n + 2m = 2(n + m)$, or $3n + 3m = 3(n + m)$, or $5n + 5m = 5(n + m)$. Similar reasoning describes the three subdivisions of numbers according to divisibility by 3. If a number has a remainder of 1 when divided by 3 it can be symbolized as $3n + 1$. If a number has a remainder of 2 when divided by 3 it can be symbolized as $3m + 2$. When we add these, we find $3n + 1 + 3m + 2 = 3(n + m + 1)$, a number divisible by 3.

Unitizing groups and operating with these groups is just the beginning step in structuring the number system multiplicatively. On the horizon (in the years to come) are related big ideas. As students discuss the structures emerging in their work, they will need to develop language and symbols to express their ideas. The symbols represent the new mental objects they are forming—all part of the developing landscape of algebra.

SUMMING UP

The children described in this chapter separated and reformulated the natural numbers as they traversed the mathematics landscape horizontally and vertically. The "connecting bonds of activity" (Dewey's phrase) that developed formed a network of relations—dense structures—that provided powerful algebraic tools for future structuring. Dewey also noted that as adults we are so familiar with these structures that it is hard for us to imagine how complex this process is. We saw that moving vertically from an additive understanding of even numbers (as $n + n$, or step by step as $n + 2$) to the multiplicative understanding of even numbers (as both $2 \times n$ and $n \times 2$) is a significant step in first grade. The notion of "threeven" extends these ideas in ways that can challenge adults!

When learners are given opportunities to organize and classify—structure—their own lived worlds, the forms they create are beautiful and surprising. The construction of Galton's "beautiful form[s] of regularity" is one of the driving forces of mathematicians. Even young children, when given the opportunity to be "young mathematicians at work," can revel in the beauty of mathematics.

4 | CONTINUING THE JOURNEY

The Role of Contexts and Models

The art of doing mathematics consists in finding that special case which contains all the germs of generality. . . . The further a mathematical theory is developed, the more harmoniously and uniformly does its construction proceed, and unsuspected relations are disclosed between hitherto separated branches of the science.

—David Hilbert (1862–1943)

The shift from additive structuring to multiplicative structuring is difficult to make. For five- and six-year-olds, using doubles and treating the group as a unit to be counted (unitizing) are landmarks on the horizon. Arriving at them is cause for celebration. Once those landmarks have been reached, however, new ones appear. A rich, dense structuring of multiplicative relations is necessary in later algebraic work, and a reliance on doubling can lead to many cognitive obstacles.

In this chapter fourth and fifth graders explore factors and multiples, unique factorization, least common multiples, and greatest common factors, as well as additive combinations related to these concepts. These forays along the vertical dimension of the landscape of learning lead them to generate and generalize the commutative, distributive, and associative properties for multiplication. Simultaneously, as they traverse the landscape horizontally, they enter the world of geometry—exploring the relationship of surface area to volume of rectangular prisms (boxes) and measuring jumps in frog races.

TEACHING AND LEARNING IN THE CLASSROOM

Miki Jensen is using the unit *The Box Factory* (Jensen and Fosnot 2007) with her fourth graders, which she introduces by showing her students a box of chocolates she received in the mail. "A friend sent me a box of chocolates the other day. I opened it and I noticed that the chocolates in the box formed an array. See, two rows and six columns." She writes 2 × 6 on the chalkboard as she talks, checking to make sure her students agree with her representation. "I discovered that there was another two-by-six layer of chocolates underneath! Look! How many chocolates are there all together?"

The students easily establish that the box holds twenty-four chocolates, so Miki writes $(2 \times 6) \times 2$, explaining that she is putting the parentheses around the 2×6 because that was the part they calculated first (the amount in the first layer) and that the $\times 2$ on the right of the expression represents that there are two layers. "This box has a layer on top arranged as two by six," she explains, "and another one just like it on the bottom: twelve in each layer, two layers, and twenty-four chocolates in the box. But then I began thinking about boxes in general—not just my chocolate box—and how they come in all different shapes and sizes. Some boxes have only one layer; others have arrangements like two by two, but then there might be many layers and the boxes are taller. If a box held twenty-four items and each layer had two rows and two columns, how many layers would there be?"

Together the class establishes that there would be six layers. Miki writes $(2 \times 2) \times 6$ and then continues. "I started to wonder about all of the possible arrangements, about box factories where boxes are made, and about how so many things come packaged like this, in rows and columns and layers. Box factories must have designers—people who decide the size and shape of boxes. What other arrangements do you think there are for twenty-four items—arrangements of rows and columns with layers? How many possible designs are there?" *One of the things mathematicians often do before they explore a problem is agree first on how to define it. They decide what the constraints will be: what will count as a solution or possibility and what won't. Here Miki supports such a conversation and the community limits the box shapes to be explored to rectangular prisms.*

The group also decides that a box rotated 90 degrees on the horizontal (or tabletop) plane (not turned on end) is really the same box and thus won't count as another possibility: $(4 \times 3) \times 2$ is the same box as $(3 \times 4) \times 2$, as it is a box with two layers, each 3×4; it has only been rotated 90 degrees. On the other hand, $(4 \times 2) \times 3$ will count as a different box because the bottom is now what was previously a side. The box has been turned on end or "flipped" as the children call it,[1] and now there are three layers of 4×2.

These fourth graders now set off to determine all the possible boxes (rectangular prisms) for twenty-four items arranged in rows, columns, and layers. Most important, they are also asked to consider how they know they have all of the possibilities. Bins of multilink cubes are available on the tables.

---

[1]The students in this class use the word "flip" for this operation, so we use it here. However, it is different from flipping a coin or a pancake, where the top becomes the bottom. One has to make sure that such socially constructed terminology is understood by the entire class and not confounded with other common uses of the word.

Context is a powerful tool that teachers can use to support mathematical development. In contrast to typical word problems found in many curriculums, which ask children only to apply what they already know, Miki is using a real-world context with two purposes in mind. She has crafted it to generate new learning, and she provides it as a framework to help children realize what they are doing. She has chosen a situation from children's lives that they can imagine, carefully picked numbers that will engender some interesting factoring possibilities, and then invited them to engage in a series of investigations during which data can be collected and examined.

At first a few students use the cubes to build rectangular prisms randomly and then count to see if they have used twenty-four cubes. Others don't include arrangements with more than one layer. But working in small groups helps to clarify the task, and soon more systematic strategies begin to emerge.

"It's, hmm, I think we need to use factors. Here's one. It's two by six—those are factors—with two layers, and now we can make a four by three with two layers," Lori explains to her partner, Michelle. "See? We can move this part around." She breaks a $(2 \times 3) \times 2$ arrangement off the original $(2 \times 6) \times 2$ arrangement and reattaches it to the bottom to make a $(4 \times 3) \times 2$.

Miki notices that Lori is using a doubling and halving strategy. To ensure that Michelle is following, she says, "Oh, that's an interesting way to find other boxes. Did you see what she did, Michelle?"

"Yeah, that's neat. Which is the bottom of your box, though? I'm confused."

Miki models how to record the arrangement. "Let's mark the bottom array with parentheses." She writes $(4 \times 3) \times 2$. "I'm just going to put parentheses around the 4×3 so we know that this is the bottom layer and the two means number of layers. Remember? Does that make sense?"

"Oh, I get it. We just doubled and halved. You broke the box in half and moved it down. Now the two becomes four and the six is three."

Miki leaves them with an important question before she moves to another group. "I wonder if this doubling and halving strategy will help you find all of the possible boxes for twenty-four items?"

Delighted with their new strategy the girls begin to work more systematically. They are beginning to consider factors and have a systematic way to make another factor pair from an original by doubling and halving, but they don't consider the number of layers as a factor as well. They work first with two layers, then with one, but ignore other possible numbers of layers. Their work is shown in Figure 4.1.

Across the room Cherise, Michael, and Aaron are working quite differently. Although they begin by using doubling and halving, as they build the boxes with cubes they realize that new arrangements can be found if they flip the box. Although the commutative and associative properties are not being used consciously and are not yet generalized ideas, they underlie these students' flipping strategy. Their work is shown in Figure 4.2.

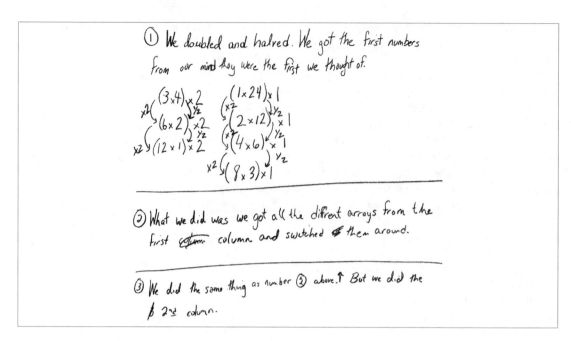

FIGURE 4.1   Lori and Michelle's Doubling and Halving Strategy

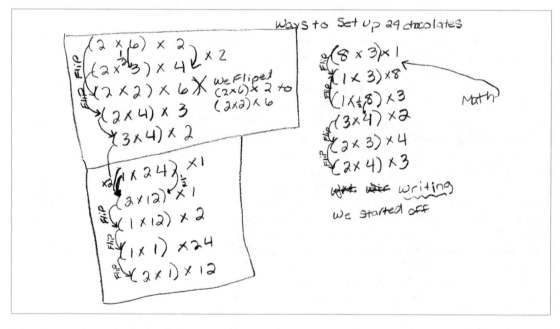

FIGURE 4.2   Cherise, Michael, and Aaron's Work

Gene, Ebon, and Brian also use a flipping strategy. Although they struggle to draw the three-dimensional boxes to scale, they have systematically used all the factors and found all the arrangements. Their work is shown in Figure 4.3.

Next Miki suggests a gallery walk to give her students an opportunity to read and think about one another's solutions. *Time for reflection is critical to learning, and a gallery walk allows students to revisit and reflect on the problem and comment on one another's mathematical thinking and representations. In this particular case, it is also a chance to examine the reasoning regarding the second question: How do we know we have found all the possibilities?* Miki passes out small pads of sticky notes and suggests that her students use them to record comments or questions and place them directly on the posters.

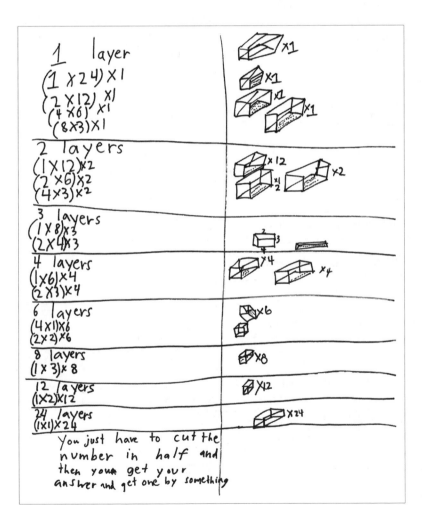

FIGURE 4.3
The Work of Gene, Ebon, and Brian

After the gallery walk, Miki convenes a math congress. Although many of the students have used a flipping strategy, they are not yet consciously aware of the properties underlying their actions. Miki wants to get her students to generalize the commutative and associative properties, so she asks Tim, Mary, and Chas, whose poster is shown in Figure 4.4, to start the discussion. Tim begins.

"First we started with all the one-layer boxes," he explains, pointing to the chart on their poster that shows boxes with one layer. "Then we halved this row." He points to the $(4 \times 6) \times 1$ box. "And then we doubled the number of layers and got another box, four by three by two."

Mary helps clarify. "How we knew that there were no more ways [to make boxes for twenty-four items] was because we started with all the ways for one layer, and then we found all the different ones for each layer up to twenty-four. So we switched the number in the rows with the number in the layers, and this is our example. We had this"—she points to $(1 \times 24) \times 1$— "and then we made it this"—she points to $(1 \times 1) \times 24$—"because we switched this number here with this number there."

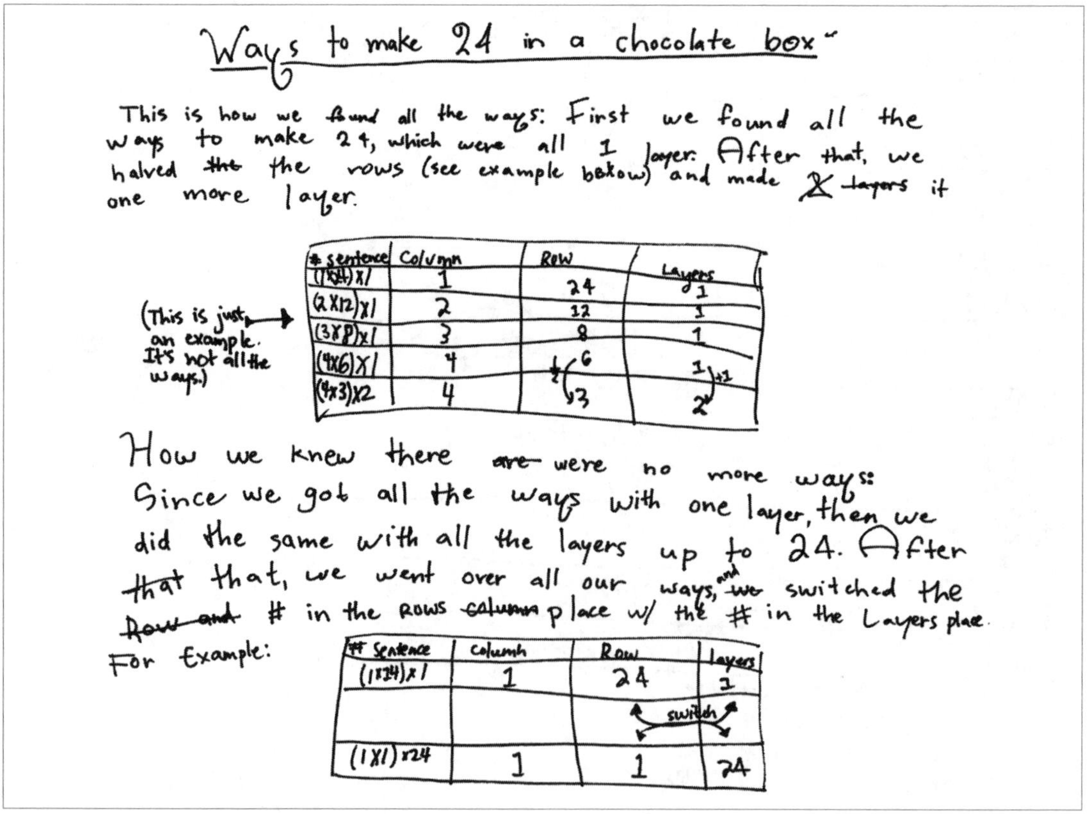

FIGURE 4.4 *Tim, Mary, and Chas' Poster*

Chas joins in. "That was kind of another strategy. We did that with all the other boxes, and later we made sure we didn't do double the same boxes."

The three students wait for questions from the class. Chas calls on Chloe, who has raised her hand.

"Right down there, I don't exactly get what you did." Chloe points to the bottom of the poster, where they have written "switched numbers."

Chas responds. "We wrote that after we began to realize that since we have all the ways for one layer, we could do the same thing for all the layers up to twenty-four. So we switched the numbers in the rows place with the numbers in the layers place to get all the ways. We had the columns stay the same, but these two"—pointing to the rows and layers—"were reversed and it's a different box."

Chloe still looks confused, and Miki attempts to focus the discussion. "Mary, Chas, and Tim, I think what some people might be confused about is what you mean by 'switched the numbers.' When you're switching numbers around, what exactly is happening with the box? If you can show us using the box, it might make it clearer for some people who are still confused."

Miki wants her students to generalize the associative and commutative properties of multiplication. Noticing the confusing language this group is using ("ways," "switching numbers," "reversed") to explain their work, she refocuses the discussion around the original context—the boxes. Staying within the context gives students a concrete tool with which to imagine the situation. In time, after they've explored several situations in which the properties appear—as they move across the landscape horizontally—they will reflect on the relationships and become more able to abstract them from the context.

"Okay, well, this is two by six by two. So it's two, six, two. *Chas holds the* $2 \times 6 \times 2$ *box made of cubes as he explains, but his language (2, 6, 2) also reflects pure number. He has already generalized beyond the context.* He points to the chart where 2 is in the column labeled *column*, 6 is in the column labeled *row*, and 2 is in the column labeled *layer*. "So this is the box flat. And then we changed it and made it tall. *He takes the same box and flips it so it is now a box with six layers and continues,* "So now it has six layers and each layer is two by two. So the six and the two were switched. Like a one-by-twenty-four-by-one is long." He stretches out his arms wide to the left and right to demonstrate and then turns the box. "But if we change it this way it becomes really tall. Then it's one-by-one on the bottom with a lot of layers, twenty-four.

Miki continues to focus on the context of the box. "How many of you found that you could turn and flip? Sometimes just turning got you the same box, didn't it? But flipping gave you a different box bottom so the parentheses changed. For example, if two by three is the bottom and you have four layers, you can move the parentheses to get a new bottom of three by four, and now there are two layers. It's still twenty-four items but now we have two layers and each layer is three by four."

Several students nod, and Katie, who has been quiet during most of the discussion, exclaims, "It's like you still have the same factors but you're grouping them differently!"

"Isn't that interesting about multiplication? We can group the factors in different ways but we still get the same answer." *Miki makes the associative property explicit.*

WHAT IS REVEALED

Miki has successfully encouraged her fourth graders to begin structuring the number system multiplicatively, but their work thus far is representative of only the very beginning of the journey. Many children have begun to consider the associative and commutative properties, but are these generalized ideas for all of them? Would they be able to generate all the possibilities in another context using the commutative and associative properties?

Just appearing on the horizon is prime factorization, where a number is factored into prime factors using exponentiation. For example, one can think of 36 as $3^2 \times 2^2$. Miki's students have a long way to go yet before this landmark is realizable. It may in fact be at least another year away. In the meantime, Miki is preparing the terrain. Underlying prime factorization is the understanding that multiples can be decomposed into factors, which in turn may be able to be further decomposed and regrouped.

BACK TO THE CLASSROOM

Over the next several days Miki begins math workshop with brief mini-lessons—strings of related problems using doubling and halving, tripling and thirding, and eventually generalizing to n and $1/_n$. She writes one problem at a time from the string below and asks for thumbs up when children have an answer:

$$3 \times 4$$
$$3 \times 8$$
$$6 \times 8$$
$$12 \times 4$$
$$24 \times 2$$
$$48 \times 1$$
$$3 \times 16$$

Then, using rectangular arrays, she represents a variety of the students' strategies for finding the products but encourages the children to make use of the relations in the string. [For more information and video of teachers doing minilessons using strings, see Dolk and Fosnot, *Multiplication and Division Minilessons* (CD-ROM), Portsmouth, NH: Heinemann, 2006.]

The first three problems are basic facts, but they are presented one at a time and related in a way that supports the noticing of doubling (each one is double the previous one). The next three problems are all equivalent, because one factor doubles while the other halves. Not all of the students use doubling and halving to produce the answers, but the fact that the answers are the same prompts a discussion of the equivalent expressions. The doubling and halving are not as easy to see in the last problem (going back to 6×8), but the students begin to look back over the string. And since the answer is the same as in the previous problems, new relationships are also examined (for example, $12 \times 4 = 3 \times 16$ and $24 \times 2 = 3 \times 16$).

Over the next several days the ideas discussed in the first string are generalized using strings like the one below:

$$6 \times 8$$
$$12 \times 4$$
$$3 \times 16$$
$$5 \times 9$$
$$15 \times 3$$
$$45 \times 1$$
$$24 \times 3$$
$$8 \times 9$$

Here, the first three problems in the string are equivalent. The first two remind students of the doubling and halving work they did earlier, but the third problem provokes some discussion. Some students note that the first problem can be doubled and halved to reproduce it, but others note that the second problem, quartered and quadrupled, produces the same result.

The next three problems invite students to consider tripling and thirding. Students point out, "Tripling and thirding is basically like doubling and halving except you have to break it up into three hunks and slide them over instead of two equal hunks." Eventually students begin to discuss the underlying reason that these strategies work—the associative property—for example $(5 \times 3) \times 3 = 5 \times (3 \times 3)$.

The next problem requires students to come up with their own helper problems. Miki uses 24 and 3 because students are quite familiar with the factors of 24 at this point. Thirding and tripling turns 24×3 into a fact they know: 8×9.

Miki also prompts her students to make use of the distributive property with this string:

$$2 \times 3$$
$$2 \times 30$$
$$4 \times 4$$
$$4 \times 40$$
$$4 \times 41$$
$$4 \times 39$$

The first and third problems are basic facts, which are related to the second and fourth problems, respectively. Each coupling supports the development and use of the associative property. For example, 2×30 can be thought of as $2 \times (3 \times 10)$ or $(2 \times 3) \times 10$. The last two problems in the string help students use the distributive property of multiplication. The 4×41 is just one more group of four when compared to the fourth problem. For the last problem, some students split 39 and calculate $(4 \times 30) + (4 \times 9)$ and add the partial products together, $120 + 36 = 156$. Other students notice a connection to the previous two problems. For example, $4 \times 39 = 4 \times (40 - 1) = (4 \times 40) - (4 \times 1) = 4 \times 4 \times 10 - 4$.

Miki also wants to encourage horizontal mathematizing and so she introduces another context for investigation. "We spent the last couple of days investigating all the different boxes for twenty-four items. We talked about how we knew we had all the possibilities, and we're confident that we've found them all. We found some boxes for twenty-four items that had pretty interesting dimensions, like the one by twenty-four by one. Imagine what this box would look like and trying to buy it at the store! Then I began to think about the amount of cardboard needed to manufacture each of the boxes. I started to wonder, would all our designs require the same amount of cardboard? If not, since it costs money to buy cardboard to make the boxes, would some of them be cheaper to manufacture? Would some be more expensive?"

THE ROLE OF CONTEXT

The first investigation required students to arrange twenty-four items in a box. By using parentheses to differentiate the rows and columns of the layers from the number of layers, students constructed and employed the associative property as they worked to find all the different ways to arrange twenty-four items (cubes) into rectangular prisms.

Building on this initial investigation, Miki now offers a context in which the students explore the surface area of these boxes and subsequently the relationship of surface area to volume. The $^3/_4$-inch multilink cube is the unit of volume; the face of the cube—a $^3/_4$-inch square—becomes the square unit used to measure the surface area. The faces of the box become smaller (less area) the closer you get to a cube and therefore require less cardboard. Eventually students will find that the closer they get to the making of a cubic prism, the cheaper the box will be to manufacture.

This new exploration is also designed to get students to realize the congruency of some of the boxes and to further support their understanding of the associative and commutative properties. For example, students notice that the cost of cardboard for a $(3 \times 8) \times 1$ box will be the same as that for a $(3 \times 1) \times 8$ and $(1 \times 8) \times 3$ box. They apply this observation to other boxes, making it a generalization. Sample students' work is shown in Figure 4.5.

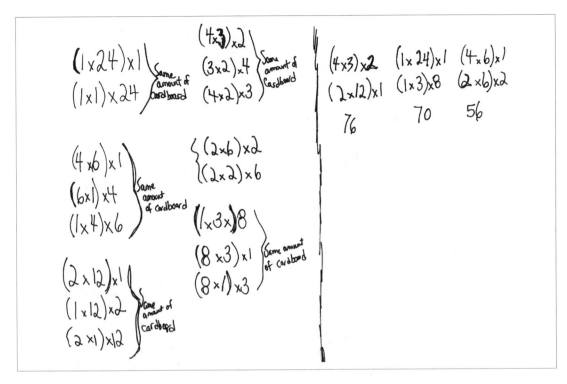

FIGURE 4.5   *Exploring Congruency*

BACK TO THE CLASSROOM

In a subsequent math congress Max and Joe discuss the relationship they have noticed about the shape of the box and the cost of the cardboard. Their work is shown in Figure 4.6.

"Well, we started like everyone else," Max begins. "We had the different boxes for the twenty-four items, and we found how much they cost and stuff. But what Joe and I are going to talk about is this." He points to the writing on their poster. "Here it says that we think the more cube-y boxes take less cardboard, so this one is cheaper." He holds up a 3 × 4 × 2 box.

Miki comments on their coined term. "Ooh, interesting word, *cube-y*. Can you talk to everyone about what you mean by it?"

"Cube-y is the opposite of long and thin. Cube-y is like this." Max holds up the 3 × 4 × 2 box.

"Why cube-y and not cube? What is the difference?"

"If the box was a cube, the length and width and the number of layers would all be the same. All the sides of the box would have to be squares.

65

We couldn't make that with the twenty-four cubes—this was the closest we could get—so we called it cube-y."

"So you're saying that the more cube-y the box gets, the cheaper it is. Why do you think that happens?"

Joe responds, referring to the work of the previous group. "Well, it's kind of like what Jeff and Maggie were saying before, about how when there's less showing it's, umm, wait . . ." He looks at Jeff and Maggie's poster still hanging on the board. "I think they were saying there's more, umm, what were you saying again?"

"The boxes that show the least squares on the outside cost the least," Maggie clarifies. "The box that shows the most squares on the outside costs the most."

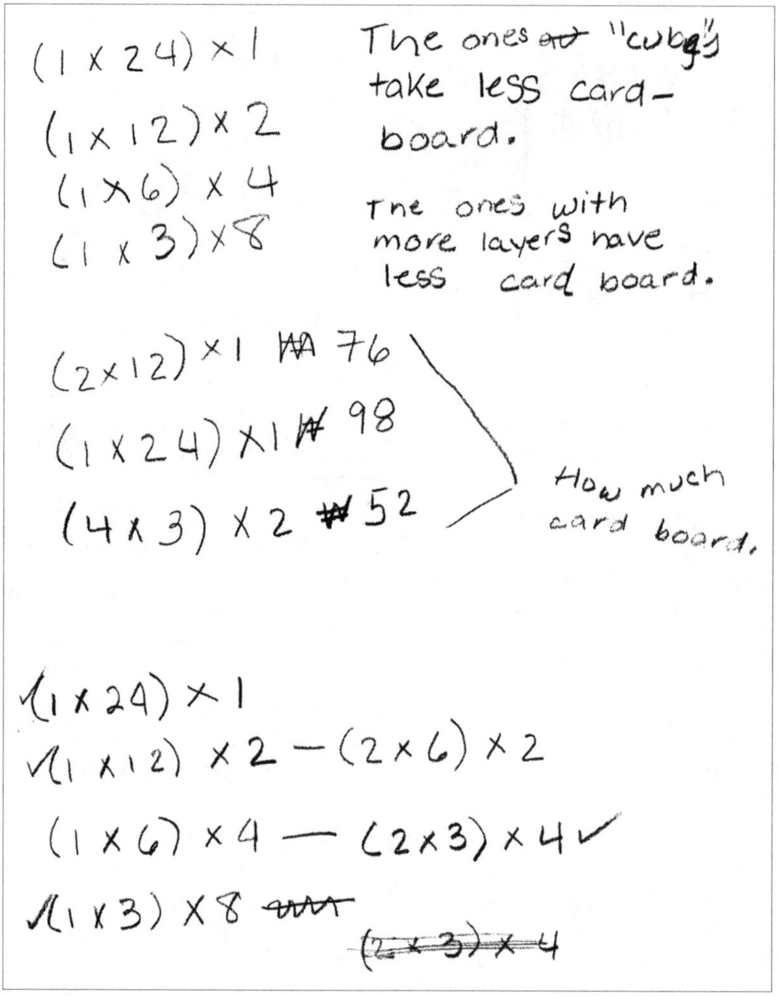

FIGURE 4.6
*The Cost of the
Cardboard: Relating
Surface Area
to Volume*

$(1 \times 24) \times 1$

$(1 \times 12) \times 2$

$(1 \times 6) \times 4$

$(1 \times 3) \times 8$

The ones are "cube's take less cardboard.

The ones with more layers have less card board.

$(2 \times 12) \times 1$ ~~AA~~ 76

$(1 \times 24) \times 1$ ~~#~~ 98

$(4 \times 3) \times 2$ ~~#~~ 52

How much card board.

$(1 \times 24) \times 1$ ✓

$(1 \times 12) \times 2 - (2 \times 6) \times 2$ ✓

$(1 \times 6) \times 4 - (2 \times 3) \times 4$ ✓

$(1 \times 3) \times 8$ ✓

~~$(2 \times 3) \times 4$~~

Joe resumes, clearer about the point he wanted to make. "Right. When the box is cube-y there's more stuff on the inside that's not showing. So you don't need cardboard for those. So it's cheaper."

Miki is elated about the way her students are imagining spatial relations about interior edges of cubes being adjacent. *This observation will motivate more multiplicative structuring as students become interested in factorizations, which may involve squares or cubes or at least be close to cubes.* "This is a pretty important point that you've noticed. Do you think that we can say, then, that the closer the box gets to a cube, the less cardboard is needed? For example, if you had eight items, what do you think the dimensions of the cheapest box would be? Turn to a neighbor and share your thinking about this question."

WHAT IS REVEALED

As Miki has encouraged multiplicative structuring, her students have journeyed vertically, developing important properties—big ideas that in later years can be represented algebraically as $ab = ba$; $(ab)c = a(bc)$; and $a(b + c) = ab + ac$. They have also traversed the landscape horizontally, applying multiplicative structuring to various geometric contexts. Being able to imagine the possible shapes of products, the factors as dimensions, the areas of the faces formed, and the congruencies of some of the shapes will be important capabilities in later algebraic work.

Because Miki's intent is to develop a rich network of relations—dense structures—there is one more part to this story. As mentioned earlier, prime factorization and exponentiation are on the horizon. A full understanding of these ideas is not a landmark goal at this point for most, but before Miki leaves the box factory exploration, she provides opportunities for these ideas to be explored as well.

THE ROLE OF CONTEXT

Miki introduces a new investigation by building on prior constructions—the idea of cubic boxes being the cheapest to make. She then offers a new challenge, asking her students to imagine three cubic boxes: a small box, $2 \times 2 \times 2$; a medium box, $3 \times 3 \times 3$; and a large box, $4 \times 4 \times 4$. She also tells them that the cost of cardboard is 12 cents per square unit and asks them to consider how many items (they can use multilink cubes) each of the boxes holds and the cost of the cardboard for each box.

The numbers for this investigation have been carefully chosen to call even more attention to the associative and commutative properties. Some students calculate the cost of each face separately; their strategy for the smallest box is represented as (12 cents $\times$ 2 $\times$ 2) $\times$ 6. Other students work with the total surface area first; their strategy is represented as (2 $\times$ 2 $\times$ 6) $\times$ 12 cents.

In the subsequent math congress, Miki addresses the idea that the factors can be grouped in a variety of ways without changing the product. As
students compare the 2×2, the 3×3, and the 4×4 faces, they also note
some interesting patterns. The dimensions (width and length) are growing
by only one each time, but the area grows quite differently. An **L** shape is
formed around the initial square; thus the increase from 2×2 to 3×3 is 5
square units; from 3×3 to 4×4, it is 7 square units. Noting these patterns
(made by the difference of consecutive squares) provides a new look at
area, arrays, and dimensions. (See Figure 4.7.) It also provides new insights
related to volume. The number of cubes held by each of the three boxes
increases dramatically, by a cubic power—from $8 = 2^3$ to $27 = 3^3$ to $64 = 4^3$.

THE IMPORTANCE OF MODELS

Woven throughout Miki's work is the use of the array model (two-dimensional
and three-dimensional). It is generated in the context of a box holding chocolates arranged in rows and columns and layers (three-dimensional), and later
it represents the surface area of the faces (two-dimensional). The array model
is a concrete tool the children can use to explore the commutative, distributive, and associative properties in relation to multiplication and to examine the
relationship between surface area and volume. Miki also uses the open array
model to represent their strategies for computation as she does strings of
related problems in minilessons.

Over time the array model will become a powerful tool for thinking—
allowing students to represent multiplicative algebraic expressions geometrically where appropriate and helpful. For example, the multiplication
$(x + 3)(x + 2) = x^2 + 2x + 3x + 6$ was traditionally taught using an acronym like
FOIL (first, outer, inner, last). Multiplying the first terms in the expressions
$x + 3$ and $x + 2$ produces x^2; multiplying the outer terms produces $2x$, the
inner terms, $3x$, and the last terms, 6. Students taught these procedures by
rote often don't understand that such a product can be represented by an area.
To make matters worse, the order of procedures they were taught for whole

FIGURE 4.7
*Comparing a
2 × 2 Square to a
3 × 3 Square*

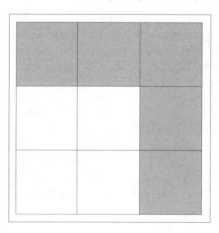

number arithmetic multiplication is the reverse—LIOF, or last, inner, outer, first! For example, imagine the same product as before, $(x + 3)(x + 2) = x^2 + 2x + 3x + 6$, but where $x = 10$. If we used FOIL, we would have $13 \times 12 = 100 + 20 + 30 + 6$. Yet the standard algorithm requires 3×2 as the first step, 3×10 as the second, 10×2 as the third, and 10×10 as the last! The partial products in both cases can be represented geometrically with arrays, and the underlying big idea is the distributive property. Students who have developed rich, dense multiplicative structures have no difficulty in understanding the relationships. But students who have been taught only rote procedures often have no image of the relationships. (See Figure 4.8.)

The number line is also a powerful representation for developing multiplicative structuring. It is well suited for exploring and representing common multiples, because they appear as the landing points when repeated jumps of equivalent lengths are taken. Factors also appear, except in reverse; they are the lengths of the jumps and the number of jumps taken to get where you want to be on the line. (See Figure 4.9.)

BACK TO THE CLASSROOM

The students in Bill's fifth-grade class, who are working on the unit *The California Frog-Jumping Contest* (Jacob and Fosnot 2007), are investigating relationships between multiples of numbers on an open number line. The context is one in which a frog and a toad jump equal distances. This context and representation will also be used to help construct the notion of variation and solving for unknowns (see Chapter 6), but at this point students are

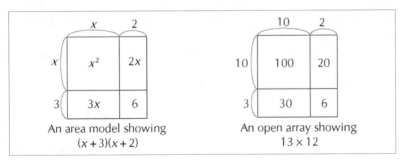

An area model showing
$(x + 3)(x + 2)$

An open array showing
13×12

FIGURE 4.8
The Area Model and the Distributive Property

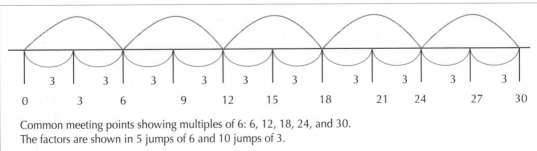

Common meeting points showing multiples of 6: 6, 12, 18, 24, and 30.
The factors are shown in 5 jumps of 6 and 10 jumps of 3.

FIGURE 4.9   *A Number Line Showing Multiplicative Structuring*

developing familiarity with the model and constructing multiplicative relations that will be crucial to what is to come. Here's the problem:

> **Frog-jump problem.** Frog jumps 8 times. Every time he jumps he travels the same number of frog steps. After 8 jumps he has traveled 96 steps. How many steps are in each of his jumps?
>
> **Toad-jump problem.** It takes Toad the same amount of time to get to 96, but he does it differently. Each of his jumps is equal to 8 frog steps. How many jumps does Toad make to get to 96 steps?
>
> **Represent both problems on one diagram** showing jumping amounts and explain how are they are different and how are they similar.
>
> **Mark the meeting points.** Where do Frog and Toad both land? Clearly, 96 is one answer. Are there other places where they might both land?

The Frog and Toad problems are placed together purposefully. Frog's problem is a form of partitive division—distributing the 96 into 8 equal jumps. The Toad problem results in the same answer (12), but it is a form of quotative division—measuring how many eights fit into 96.

Tom and Alicia have just finished presenting two diagrams illustrating their solutions to the Frog and Toad problems. Most students have solved it rather readily—a few students writing as little as "96 ÷ 8 = 12" and stating, "This is easy!"

Bill asks Tom and Alicia, "Okay, you have two nice diagrams here, and you found that Frog jumps 12 steps each time and Toad takes 12 jumps. Nobody asked you any questions, which is unusual for this group. You made two diagrams. Let's make one together and put both Toad's and Frog's jumps on the same line." Tom, with Alicia's help, makes a new drawing shown in Figure 4.10. "You started carefully, but then you drew the last jumps quickly. Why?"

Tom replies, "We realized it was just the same pattern over again so we did it quickly."

"What do you mean? What pattern?" asks a classmate, Sylvia.

"See these three jumps of Toad's?" Tom explains. "They are like two jumps of Frog's. That pattern goes over and over again. In fact it would keep on going if they didn't have to stop at 96."

"Tom, show us these chunks, or patterns as you call them. Circle them in green and let's think about how many there are and how big they are." *Proportionality is an important big idea so Bill asks for clarification in an attempt to help more students understand Tom's insight.*

"Well, there are four of them, and each is 24. I guess that's because three 8s is two 12s. There would be more than four if they didn't have to stop at 96."

Alicia records what Tom has said: "3 × 8 = 2 × 12, 4 chunks of 24." Then she adds, "96 ÷ 8 = 12, 96 ÷ 12 = 8."

Bill suggests, "We have two ways to see 8 × 12 = 96 on this diagram. And we can think of Alicia's four chunks of 24 as 4 × 24 = 96. Talk with your partner about how all these numbers are related. We got 96 twice. Think about the context. Do you think this is a coincidence?"

Bill moves around the class, conferring with groups as they work. Many times he has to ask them to explain where the 12 is. The 8s are clear from the context, but the 12s are not as easy. But soon students see that the 8 and 12 are switching roles. For Frog, there are 8 jumps of length 12; for Toad, there are 12 jumps of length 8. A number of groups are also noticing that 24 is a common multiple of 8 and 12. In fact, it is the smallest common multiple, or least common multiple. Bill introduces the least common multiple notation: 24 = LCM(8,12).

The 4 remains a mystery to most of the students. There are four groups of the pattern that Tom reported initially, but how does it relate to the original numbers? While conferring with her partner, Samantha has an insight. "You know, if you think of each of these meeting places as a jump, then there are four meeting jumps inside Frog's eight jumps, and there are four meeting jumps inside Toad's twelve jumps. It has to divide both—it's the biggest factor of both."

Overhearing, Bill asks her to explain it again. He introduces the greatest common factor notation, 4 = GCF(8,12), and asks Samantha's group to think about the relationship between LCM and GCF in this context and to make a poster, a recording of their ideas, which they will present in a math congress the next day.

WHAT IS REVEALED

Earlier, some students were reluctant to create a diagram representing the two forms of division on the number line. The division problem was easy and they didn't see the point: They could easily get the answer. But Bill's request for representation on the same line pushed them to continue to investigate and moved them beyond an initial disinterest. The mystery of the 4 and 24 multiplying to 96 is a key to unlocking the important relationships between LCM and GCF. Bill is guided by his knowledge of the landscape here: There is some very interesting mathematics embedded in the merging of these two diagrams.

The problem also offers room to differentiate instruction. Some students may not have yet constructed the relationship between partitive and quotative division. Here they may notice that these two forms of division are related to "what you are multiplying when you turn it around"; that is, are you thinking of 12 jumps of 8 steps or 8 jumps of 12 steps. They are the same because of

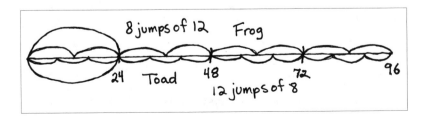

FIGURE 4.10
Tom and Alicia's Double Number Line

the commutative law: $12 \times 8 = 8 \times 12$. The commutative law of multiplication is linked with different forms of division! For many students, this is an important observation stemming from the pairing of these two problems.

For other students, the model is a way to understand the least common multiple, and how it is related to other common multiples. The common multiples are where Frog and Toad both land. This is why students are asked to find the places where they meet. Most note what Tom and Alicia shared with the class—that the pattern repeats (showing that every common multiple of 8 and 12 is a multiple of the least common multiple). In other words, because LCM(8,12) = 24, all the common multiples of 8 and 12 are 24, 48, 72, 96, continuing in increments of 24. Tom knew you could go beyond 96, because he mentions that there would be more if Frog and Toad didn't have to stop at 96.

The work with greatest common factors is more challenging. However, as Samantha notices, if a number of equal jumps fits inside a larger number of different-size (but smaller) equal jumps, then the first number will divide the second. In the example, 4 divides both 8 and 12. This is a tricky transition in this context, because one is now thinking about the number of jumps rather than the length of the jumps. But the students who find this can now merge their observations. The original $8 \times 12 = 96$ can be merged with the fact that the four chunks of 24 produces 96 too, because they come from the subdivision of the original 96. Now, 4 = GCF(8,12) and 24 = LCM(8,12).

Consider the factors of 5 and 10 ($m = 5$, $n = 10$) as jumps on the number line. The greatest common factor is 5, as 2 jumps of 5—or 5 jumps of 2—result in 10. The meeting points are at multiples of 10; thus 10 is the least common multiple. And 5×10 (GCF × LCM) = 50. In later years students will explore the formal proof that $mn = $ GCF$(n,m) \times $ LCM(n,m), but for now they have a model that can provide some generalizable ideas that could lead to a proof.[2] Most important, the students are constructing multiplicative relations that will certainly move them toward greater success in algebra.

WHAT MODELS ARE IMPORTANT FOR MULTIPLICATIVE THINKING?

Current cognitive scientists believe that number and space are linked—that our brains are wired to understand number and space together (Dehaene 1997, Izard et al. 2008). Using models like arrays and number lines builds on our natural capacity to intuit number relations spatially.

The models allow students to treat numerosities as the basic mental objects that are formed by direct experience—using the *common sense*

---

[2]Most proofs of this result make use of unique factorization, a big idea toward which these students are en route but which is still on the horizon for many. The spatial approach given in this instance is generalizable—its most tricky aspect is recognizing the GCF in the model.

approach is rooted in the use of unique factorization and makes use of pre-formed mathematical structures defined by pure number relations devoid of context, namely the existence of a factorization into primes, and breaking these apart to find the relationship between GCF and LCM. Because of the axiomatic nature of the reasoning, it applies more generally (to polynomials for example), but this likely means that approaches like this should come much later in students' development.

For example, many textbooks teach students to build factor trees. This approach stems from the pedagogical belief that teaching this process and how to represent it will lead to understanding. In our experience, many students who build these trees have either lost sight of or never actually constructed the multiplicative relations that underlie the process. They can carry out this procedure successfully but cannot use the unique factorization that results from the process, saying they have "forgotten what to do." The uniqueness properties may not be in sight either. Some students who build a factor tree for 32 will, even after finding all the 2s in the tree, be uncertain whether 3 could be a factor. In fact, factors of a number come from a pair of numbers—this is the multiplicative relation that defines them, and it is at the heart of the partitive and quotative structures discussed earlier. It is also at the heart of the array model in which the factors of the total area are the measures of the sides.

A number has a set of factors, and this mental object needs to be acted upon using certain operations. The pairing within the set of factors is an important piece of the structure, and it is by operating with this pairing that many algebraic constructs about number become accessible. Some curriculums have students build what they call *rainbow diagrams*, where the factors of a number are listed in ascending order and the pairing between factors is indicated by an arc, the collection of which form the rainbow. This is a powerful representation, but it is critical that students first create a mental object based on meaningful interactions within a context (the number line, for example), after which this particular representation, or another, may make sense.

The creation of a mental object (which the learner may represent in many ways) provides an opportunity to access many algebraic ideas within number. Among these are primes, common factors and multiples, and unique factorization, as well as access to more robust strategies within arithmetic operations. Dense multiplicative structures are the result of becoming intimately familiar with the sets of factors of many numbers and using them to build other sets of factors, all of which are bound up with the development of fluency with facts and the development of what we call *multiplicative number sense*. Students with good multiplicative number sense will, when asked to discuss 49, not only say, "Forty-nine is one less than 50" (an additive understanding), but will also note, "It is 7 squared, so its only factors are 1, 7, and 49." In contrast, 48 is full of factors—{1,2,3,4,6,8,12,16,24,48}. Forty-eight and 49 differ by one, but their multiplicative behavior is very different! The problems students are given must allow them to build these mental objects. Not only do these structures provide access to algebraic ideas within number, but as students generalize their features, they will also provide access to related ideas in symbolic algebra as well.

MORE EXAMPLES: LINEAR COMBINATIONS

There are rich algebraic structures associated with adding multiples of different numbers. Problems of this type provide contexts for further multiplicative structuring and opportunities for students to increase their repertoire of multiplicative relations between numbers, sets of factors, multiples, and factorizations. These can surface in a variety of problems and are often presented in a postage-stamp context—for example, finding the values that can be made using two types of stamps of given values. Investigating sums of multiples of three- and six-cent stamps (or of six- and fifteen-cent stamps) should lead one to inquire about uses of the distributive law. In effect, sums of multiples of 3 are always multiples of 3. Ask yourself, or a student, "Is 27 + 99 divisible by 3?" If this instance of the distributive law is interpreted within a rich multiplicative structure and with a vision of what that might look like on a number line, one would not have to calculate to see that the answer is yes. This is a basic facet of multiplicative structuring and can be represented on a number line in repeated jumps of threes (9 threes and then 33 threes for a total of 42 threes), or with open arrays $(3 \times 9 + 3 \times 33)$. Such mental spatial objects for the distributive law underlie the development of many structures critical to algebra.

Adding multiples of numbers without common factors also leads to important algebraic structures within number. Adding a four-cent stamp to an inventory of six-cent and fifteen-cent stamps enables the possibilities of forming 18-, 19-, 20- and 21-cent combinations, and therefore all larger combinations (17 is not possible). This is a result of a big idea known as *cyclicity*. Once one has the numbers 18, 19, 20, and 21 as sums, one can add multiples of four to these and, cycling through, obtain all larger numbers. The mental object upon which this big idea is built is the listing of possible remainders when dividing by four. The remainders are 0, 1, 2, and 3. This structure can be derived when divisibility by 4 is studied and builds on the partition of whole numbers into even and odd (see Chapter 3). Students need to construct similar partitions for divisibility by 3 (into three classes), as illustrated earlier in our discussion of "threeven" numbers. These mental structures can then be represented and generalized. Postage-stamp problems are one context in which acting with these objects is a powerful problem-solving tool—far more robust than guess-and-check methods.

Unfortunately, many problems presented to students don't produce the structures and mental objects we want to encourage. For example, Lager (2007) has written about a professional development session in which teachers considered a problem involving possible change received from $5.00 after purchasing 150 three- or six-cent stamps, the exact number of each stamp being unknown. The problem posed was, "Would $.74 be possible as the correct change?" In a course Bill co-taught with Lager, university students worked on the problem and then studied his article. Interestingly, after considerable discussion of language issues and problem-solving approaches (including various facets of guess-and-check approaches) and examination of work by teachers and students, the fact that whatever amount was spent had to be a multiple of three rarely surfaced. This fundamental consequence of the structure of sums of multiples was not on their radar screen. Students' failure to grasp such a basic idea is something all mathematics educators need to work on!

SUMMING UP

It is impossible to talk about mathematizing without talking about modeling. Mathematical models are mental maps that depict relationships helping us to understand and represent our world. As such they are powerful tools when solving problems. Models themselves are constructed. They emerge from representations of the *action* in the situation. Later these representations develop into *representations of the situation* using cubes or drawings. Eventually, modeling develops into a symbolic *representation of the mathematizing itself and becomes a tool to think with*.

In this chapter we witnessed fourth graders decompose numbers into their factors and generate the commutative, distributive, and associative properties as they pertain to multiplication. Along the way, a rich set of geometrical representations also developed building on students' intuitive natural abilities to unite space and number. Multiplicative relations can also be represented powerfully on a number line. We saw fifth-grade students representing least common multiples in a context where the representations of two problems, one involving a frog jumping and the other a toad, were merged onto a single diagram. Again, spatial modeling was key to multiplicative structuring.

The movement from structuring the number system additively to structuring it multiplicatively is an important developmental step in the algebra landscape. In order to develop a dense collection of multiplicative structures, students need repeated opportunities to construct mental objects of sets of factors, including information about primes, common factors and multiples, and unique factorization, all of which are bound up with the development of fluency with multiplication and the development of what we call *multiplicative number sense*. Contexts that lead to array and number line models play a crucial role in this development.

As David Hilbert says in the epigraph to this chapter, "The further a mathematical theory is developed, the more harmoniously and uniformly does its construction proceed, and unsuspected relations are disclosed between hitherto separated branches of the science." The relationships students found as they examined the special cases related to the boxes and frog jumping helped to foster the development of generalities and the development of a rich, dense multiplicative structure—one that will become a powerful lens in their later algebraic work.

5 | EQUIVALENCE ON THE HORIZON

Everything should be made as simple as possible, but not simpler.
—Albert Einstein

It seems that if one is working from the point of view of getting beauty in one's equations, and if one has really a sound insight, one is on a sure line of progress.
—Paul Dirac

Equivalence—knowing that two expressions may be equivalent even though they don't look alike—is one of the most important ideas in early algebra. Equations are based on this concept. Unfortunately, young learners often think the equals sign just means "the answer is coming" (Carpenter, Franke, and Levi 2003), probably because many teachers tend to write equations depicting the result of arithmetic procedures, with the answer to the right of the sign. This misconception may also be reinforced by the use of the = button on a calculator.

Early in the development of mathematical understanding children struggle to understand how statements such as $5 + 3 = 4 + 4$ can represent equality—particularly when they are presented only with the symbols. "The numerals are different, so how can the two sides be the same?" they ask, unable to differentiate equivalence from being identical. Thus they need to add the numbers on each side; only after achieving $8 = 8$, are they certain. As they construct the idea of compensation—in this case, that one is removed from the five but added to the three—they begin to understand how things can be equivalent without necessarily being identical.

Early number sense and the ability to judge magnitude and equivalence may be directly connected to spatial sense. It is easier to determine magnitude when the difference between the numbers chosen for comparison is greater, presumably because we can rely on our spatial sense. When numbers are close together, we shift to a mental counting strategy, an activity that researchers now think takes place in a different part of the brain than the area used for spatial/number sense. For example, if you are asked to choose which is greater, 2 or 9, you would respond 9 in a split second, and one might think comparing 7 to 8 could be done equally quickly. Surprisingly, the comparison when the numerals are close together requires a longer response time (Dehaene 1997).

Researchers also now believe that the brain is hardwired to perceive (subitize) small amounts without needing to count one by one. For example, infants can tell the difference between two and three objects and can do early addition and subtraction—they know when one is missing (Wynn 1998). The same mathematical ability appears to be true even of raccoons, dolphins, and monkeys! To make use of this natural capacity for comparing and examining equivalence, and to build on it, we often flash images of small amounts (3 and 2 on one arithmetic rack and 4 and 1 on another, for example, or 3 and 3 and 3 and 2), and ask children to determine which is more or whether they are equivalent.

Grounding the work in realistic contexts from children's lives also helps them realize that amounts can look different but still be equivalent. In the children's book *The Sleepover* (Fosnot 2007c), eight children rearrange themselves on a pair of bunk beds to confuse the babysitter, who brings up eight cups of popcorn (two groups of four cups each—four for the kids on the top bunk and four for the kids on the bottom bunk) only to find five children on the top bunk and three on the bottom. She thinks she has gained a kid and is amazed to find that four cups and four cups is perfect for serving five kids and three kids! She then goes down to get juice and brings back glasses in groups of five and three only to find that the children have now rearranged themselves into groups of six and two. After reading this story to children, you can ask them to find all the ways the kids can rearrange themselves on the bunk beds to trick the babysitter. The early development of equivalence and compensation will be seen in exclamations like, "It is still eight! One kid just went up the ladder!"

Games provide young children with further experiences with equivalence and help them connect the written symbols with the perceived amounts (see Fosnot and Cameron 2007 and Fosnot 2007a). For example, children can play bunk beds in pairs, taking turns rolling a pair of dice. If a 5 and a 1 are rolled, player 1 takes 6 counters and places them on the game board on two lines, 5 on the top line and 1 on the bottom line, and writes 5 + 1. Player 2 then draws a game card (which either says, "1 up the ladder" or "1 down the ladder") and arranges counters on his side of the game board according to the card's instructions. For example, if the card says "1 down the ladder," player 2 puts four counters on the top line and two on the bottom, records 4 + 2, and adds the equal sign to complete the equation: 5 + 1 = 4 + 2. (See Figure 5.1.)

Another game that supports the development of equivalence is part/whole bingo. Players take turns rolling a pair of dice and placing that amount of connecting cubes onto tracks on their game cards. The tracks they cover do not have to be the same as long as the total number of cubes is used. For example, if a 5 and a 2 are rolled, one child might place 7 cubes on the 7 track, but the other child might cover the 5 and 2 tracks or the 3 and 4. It is also possible to cover tracks 2, 2, and 3 or tracks 6 and 1, and so on. As long as only 7 connecting cubes in total and no partial tracks are used, the cubes can be arranged in any way the players want. The objective is to eventually cover the entire game card. (As with bingo, each game card is different.) Play is cooperative rather than competitive; players are encouraged to help each other. The game ends when both game cards are covered. The point of the game is to make equivalent amounts rather than simply finding and covering the sum.

The capture ten game supports the "making-ten" strategy for automatizing the basic facts (for example, 9 + 6 is much easier to recall when a child thinks of it as being equivalent to 10 + 5). Again, two children play together. A deck of playing cards with the face cards removed is placed in the center with cards stacked facedown. The game board has ten pockets labeled with the expressions 10 + 1 through 10 + 10 (see Figure 5.2). Each player turns over a card. Together players determine in which pocket to place the cards. For example, if 8 and 7 were drawn they would be placed in the 10 + 5 pocket. (They might be asked to justify their thinking, perhaps using the arithmetic rack.) If the sum of the two cards is less than 10 (5 and 2, for example), players put the cards back in the deck and reshuffle.

Games like these are just the beginning steps in understanding equivalence. In grades 2 and 3, students can analyze statements like 8 + 2 + 10 = 12 + 4 + 4, as well as make the generalization that equivalence is maintained no matter what operation you use on an equation, as long as you do it to both sides (thus the statement n + 8 + 2 + 10 = 12 + 4 + 4 + n is also true).

Initially children approach equations like this arithmetically; they proceed with the necessary operations left to right. To determine whether the statement 8 + 2 + 26 + 2 = 28 + 8 + 2 is true, they compute 8 + 2, add 26 next, and then add 2. On the right of the equation they start with 28, add 8, and then add 2, producing 38 = 38. One of the big ideas on the landscape now is for children to treat the expressions as objects that can be operated on—to prove equivalence *without* doing all the arithmetic. For example, if two of the numbers to the left of the equals sign are combined (26 + 2), the result is 8 + 2 + 28 on each side. Alternatively, children might

FIGURE 5.1
Playing Bunk Beds

subtract 26 from each side and produce 8 + 2 + 2 on each side. Done! No need for further arithmetic.

When students encounter expressions with variables in later years, a major difficulty is that they attempt to use arithmetic and when it doesn't work they don't know how to proceed. How do you interpret $2x + 5 = 3x - 6$ if you don't know what x is? But when you understand that the expressions are objects and that amounts can be added to or subtracted from them and equivalence is maintained, you just add $-2x + 6$ to both sides, resulting in $11 = x$. Children can learn these procedures to solve for unknowns, but if they have not constructed the idea of expressions as objects that can be operated on, the procedures are rote and without meaning.

There are several ways to operate on expressions. Using the associative and commutative properties is one way. For example, the problem $19 + 32 + 8 = ?$ can be solved by adding left to right. But the problem can be made a bit easier to solve by employing the associative property and grouping the 32 and 8 first: $(19 + 32) + 8 = 19 + (32 + 8)$. Children may understand that they can group numbers in addition without changing the total sum, but they often don't think to use this strategy. Besides being an important big idea to develop, the associative property of addition can be helpful to simplify equations because the grouping may allow for "canceling" equivalent expressions on each side of the equals sign. The same can be said for commutativity. When analyzing the statement $8 + 15 + 3 + 2 = 15 + 8 + 5$, using a combination of the associative and commutative properties allows children to conclude that the statement is true without doing all the arithmetic to derive $28 = 28$.

Algebraically, commutativity can be represented as $a + b = b + a$ and associativity as $(a + b) + c = a + (b + c)$. Children don't need to be intro-

FIGURE 5.2
Playing Capture Ten

duced to formal notation like this at this point, but they do need many opportunities to compose and decompose numbers and generalize that numbers can be grouped in a variety of ways, even turned around, and the amounts stay the same. (These properties do not hold for subtraction, and children may be amazed and puzzled as they explore this difference.)

Once children understand these ideas, they become willing to abandon their earlier, more tedious, arithmetic strategies. They begin to substitute one expression for another. For example to subtract 8, a child might subtract 10 and then add 2, or she might solve $32 + 38$ by substituting the equivalent expression of $30 + 40$. As their understanding of equivalent expressions increases, children may begin to simplify equations by eliminating quantities that are on both sides of the equals sign. For example, in the statement $8 + 2 + 26 + 2 = 28 + 8 + 2$, the $8 + 2$ can be "canceled" on each side to produce $26 + 2 = 28$. At first children draw lines through the identical numbers on each side of the equals sign but won't erase them, or they talk about ignoring or separating off those numbers momentarily. As they become more confident that the equivalence remains, they develop what they often call a "cross-out" rule.

Once children are confident that quantities can be "canceled" and that adding or subtracting identical quantities on each side of an equation won't disturb equivalence, they develop an undoing strategy for simplifying equations. For example, they might add 6 to each side of an equation to get rid of –6, or subtract 6 from each side to get rid of +6.

Developing these ideas and strategies is a huge leap in development and won't happen by just working on procedures and skills. It takes conscious thought on the part of the teacher to ensure that progress occurs toward these landmark ideas and strategies.

TEACHING AND LEARNING IN THE CLASSROOM

Cynthia Lowry is using the unit *Trades, Jumps, and Stops* (Fosnot and Lent 2007) with her second graders. She begins by reading the story *The Masloppy Family Goes to New York City* as a context for investigating how to divide the money from a giant piggy bank (6 rolls of quarters, 3 rolls of dimes, 3 rolls of nickels, 3 rolls of pennies, 4 loose quarters, 7 loose nickels, 1 loose dime, and 5 loose pennies) into three equivalent piles. The numbers of rolls (six and three) have been chosen to encourage children to distribute them easily into three piles. The real problem lies in finding ways to distribute the loose coins (which add up to $1.50) equally.

If children begin by distributing the rolls into three equal piles, they usually do the same with the four loose quarters, the seven nickels, and the five pennies. The constraint of what to do with the remaining quarter, dime, nickel, and two pennies requires that they exchange equivalent amounts. In the following excerpt from a conference with Jasmine and Keshawn, Cynthia encourages them to deal with equivalence and exchange.

"I think we need to add up all this money first," Jasmine asserts as she and her partner, Keshawn, begin.

"That's a lot of money. Two, four, six. Six dollars for the nickels! Write that down," Keshawn tells Jasmine, recognizing they need to keep track of the amounts. "Six rolls of quarters—10, 20 . . . 60. Sixty dollars in quarters. Wow! That's a lot of quarters!"

"It is, isn't it!" Cynthia says. "I wonder if we need to add all this up. Could we divide it without adding it all up?" *By pondering aloud whether addition is needed, Cynthia invites consideration of an alternative strategy. She does not suggest that the rolls be shared as wholes; she only wonders aloud whether another way is possible. She respects the children's autonomy and trusts them to generate clever solutions.*

Jasmine rises to the challenge. "I could pass out the rolls. One, two, three. . . . One, two, three. . . ." Next she and Keshawn will be faced with how to distribute the loose coins.

"Passing everything out to three piles—that's a great idea. I'll check back with you in a little bit."

During the investigation, Cynthia is only able to confer with four or five pairs of children, but they discuss and reflect on the problem with each other, working and learning autonomously. As Cynthia observes the children, she notes that several big ideas about algebra are emerging: equivalence and substitution (exchange); the generalization that a substitution strategy opens up possibilities; and the idea that the rolls are not important because each pile has the same number, so the most important part of the problem is how to distribute the coins to make equivalent amounts. After the students have had enough time to explore the problem thoroughly, Cynthia convenes a math congress to focus on these ideas.

"Ian, would you and Peter begin? Tell us what you did." *Starting with children who distributed and then were puzzled about what to do next, Cynthia frames the main problem that will engender the strategy of substitution and a discussion of equivalence.*

"We shared out all the rolls first. That was easy," Ian declares. "Then we shared out the coins. We gave every pile a quarter. Then we did two nickels each. Last we did the pennies. But then we had this left over. The boys hold out one quarter, one dime, one nickel, and two pennies. We're stuck. It's hard."

"This is tough, isn't it?" Cynthia agrees sympathetically. *Acknowledging that the problem is difficult allows the children to feel comfortable admitting they aren't sure what to do. It also lets them take risks and work as a community to help one another. Too often as teachers we think we should eliminate struggle; we offer our strategies too readily or simplify problems into trivialized tasks to ensure easier answers. Doing so unfortunately eliminates all opportunity for children to learn the tenacity needed to work on mathematics and to feel the resulting exhilaration of solution when grappling with challenging problems.* "It seems that sharing the rolls was the easy part. I guess we don't have to worry now about the rolls." *Cynthia's comment that they don't have to worry about the rolls helps the children see how equivalent amounts can momentarily be disregarded. Over time this will result in a "canceling" strategy.* Since the emphasis is on proving equivalence, she continues, "But what should we do with the loose coins? Does anyone have a way to solve this?"

"I do," Jasmine says. "You can trade. Put a dime for two nickels. Then we can give two piles another nickel and the other pile can have five pennies. That's fair."

Realizing that the children need time to think about this, Cynthia suggests a few moments of pair talk. "Turn to the person next to you and discuss what Jasmine did. Is this fair?"

After a minute or two, Amirah voices the question many of these young mathematicians have. "Yes, we agree. But we still have a quarter and a nickel left. What do we do with those?" *Although Jasmine's idea to use equivalence is a brilliant insight, her specific choice of coins does not solve the problem.*

"Did anyone find a way that worked? Is there a different exchange we could do?"

"We found a way," Carmen says excitedly. "José and I gave two quarters for one pile, and two quarters for another. Then the third pile had all of the other coins."

Cynthia says, "Let me write that down so we can all consider it." She writes Rolls + 2$\textcircled{25}$ = Rolls + 2$\textcircled{25}$ = Rolls + 7$\textcircled{5}$ + 5$\textcircled{1}$ + 1$\textcircled{10}$. "What do you think? Does this work?"

Sally, who has been looking perplexed until now, exclaims, "It works! Seven nickels and five pennies and one dime is the same as two quarters! Each pile has rolls and fifty cents!"

"Yep," her partner, Marcus, chimes in, "and there's another way, too."

"Another way?" Cynthia smiles encouragingly. "Let's check it out." Then she challenges them again. "I wonder how many ways there are to do this?" *By exploring alternative solutions, Cynthia establishes that there are other possibilities and encourages her young mathematicians to inquire further. An important part of algebra is generalizing and proving that all possibilities have been found.*

WHAT IS REVEALED

Cynthia has been using the Masloppy's trip to New York City, which necessitates that the money in the piggy bank be equitably distributed into three bags, as a context in which to develop early algebra ideas related to equivalence, substitution, and cancellation. The rolls of coins support simplifying the problem by separating out equal amounts, and the loose coins prompt the students to use equivalent amounts rather than combinations of identical coins.

A realistic context like this not only helps children realize what they are doing and makes math meaningful but it can also be specifically crafted to support development. Sequencing these activities progressively is critical. Powerful learning only results when ideas are explored in depth over time—introduced through hands-on activities and then revisited perhaps weeks or months later in different ways. Without strong *sequences* of carefully crafted activities, learners may have insufficient opportunities to develop the dense structuring important for later algebra.

At this point, Cynthia is just starting the journey of developing equivalence. The ideas are not yet explicit nor generalizable, but the terrain is being prepared. This early work with equivalence sets the stage for establishing the use of equivalent expressions and for analyzing equations. Over the next several weeks, Cynthia continues to work on these ideas progressively with a sequence of investigations, games, and minilessons.

For example, she plays twenty questions, inviting children to figure out what coins, which total fifty cents, she has in her hand. She also introduces the piggy bank game (see Figure 5.3). Two children, in turn, select a card from the game deck that indicates an amount of money they can place on their game card. Players determine what combination of coins to take. For example if a "15 cents" card is drawn, the player can take fifteen pennies, or a dime and a nickel, or three nickels, and so on. These coins are then placed on the player's board and the transaction is recorded on a

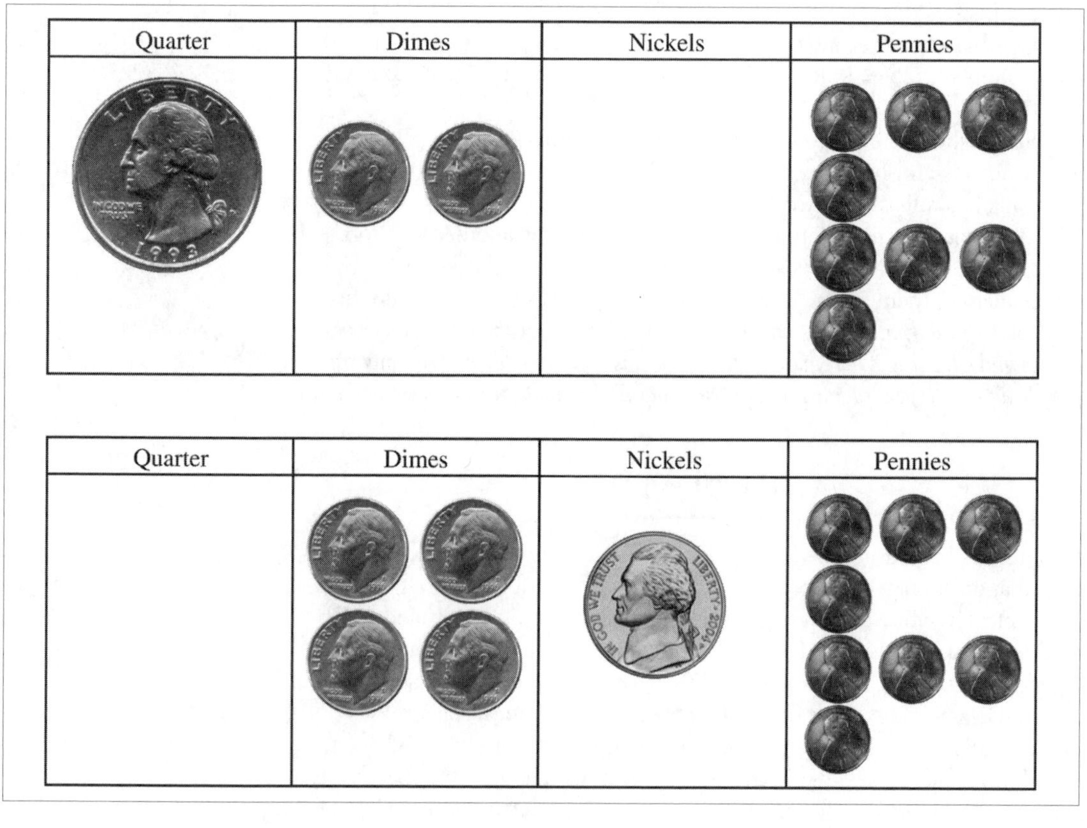

Quarter	Dimes	Nickels	Pennies

Quarter	Dimes	Nickels	Pennies

FIGURE 5.3   *Sample Game Boards*

shared recording sheet. Each pair of transactions is then linked with <, >, or =, whichever symbol is appropriate.

BACK TO THE CLASSROOM

Cynthia is conferring with Philip and Isaac as they play piggy bank.

"I have 41 cents," Philip declares, pointing to his board, which has one penny, one nickel, one dime, and one quarter.

"How do you know?" Cynthia asks.

"I added it up." Philip easily adds up the fives and tens. "A quarter is 25 plus 10 . . . 35 . . . plus 5 . . . 40. So it's 41."

"What about you, Isaac?" On his board he has three pennies, one nickel, one dime, and one quarter.

Adding up all the coins is difficult for Isaac, so he estimates. "I think maybe 47."

"Do you have more or less than Philip? Let's look at the boards together. Is there another way to tell without adding it all up?" *Cynthia prompts Isaac to use a spatial sense of number, to compare equivalent amounts, and eliminate unnecessary information.*

"He has more, I know. Wait, maybe I do. Because we both have this." Isaac places his hand on top of the nickels, dimes, and quarters. But I have three pennies!"

"So who has more?" Cynthia asks.

"I do! I guess I have 43 cents, because I have two more pennies than him!"

Encouraging Isaac to examine the expression without doing arithmetic challenged him to stop guessing—to consider equivalent expressions.

Cynthia helps Isaac record the amounts and relate them to Philip's amounts using the appropriate symbols (see Figure 5.4).

Later, Isaac plays a round of piggy bank with Olivia and teaches her the strategy he developed when he was playing with Philip.

After the students have played several rounds of the game with different partners, Cynthia convenes a math congress. "Olivia and Isaac, tell us about the strategy you found helpful."

Isaac begins, "I had three nickels and two quarters and two pennies. Olivia had two nickels, three dimes, one quarter, and two pennies. We decided it was equal."

Cynthia writes $3(5) + 2(25) + 2(1) = 2(5) + 3(10) + 1(25) + 2(1)$.

Olivia explains, "We knew two dimes and a nickel made a quarter. So Isaac traded."

Cynthia represents their thinking as an emergent two-column proof:

$$3(5) + 2(25) + 2(1) \, ? \, 2(5) + 3(10) + 1(25) + 2(1)$$
$$2(10) + 1(5) = 1(25) \, \text{(Trade)}$$
$$3(5) + 2(25) + 2(1) \, ? \, 1(25) + 1(5) + 1(10) + 1(25) + 2(1) \, \text{(Equivalence)}$$

Isaac continues, "And we both had two pennies and two quarters so we knew that part was the same and so they didn't matter."

"Wow. You said a lot there. Let's see what everyone thinks. Isaac said that the two pennies and the two quarters don't matter because they are on both sides. Turn to the person next to you and talk about this. Is he right?" *Cynthia encourages the community to consider Isaac's justification.*

Mia says, "He's right. They are the same."

"So can I erase them?" Cynthia inquires.

"I don't think you can erase them."

"Cross 'em out," Isaac says. "You could do that because we both have them. They don't matter. And a dime is two nickels. So, see, they are equal. So we put an equal sign."

"Does everybody agree that if the same thing is on both sides we can cross them out?" *Cynthia is establishing a "cancellation" law relative to equiva-*

FIGURE 5.4
Isaac and Philip's Piggy Bank Recording Sheet

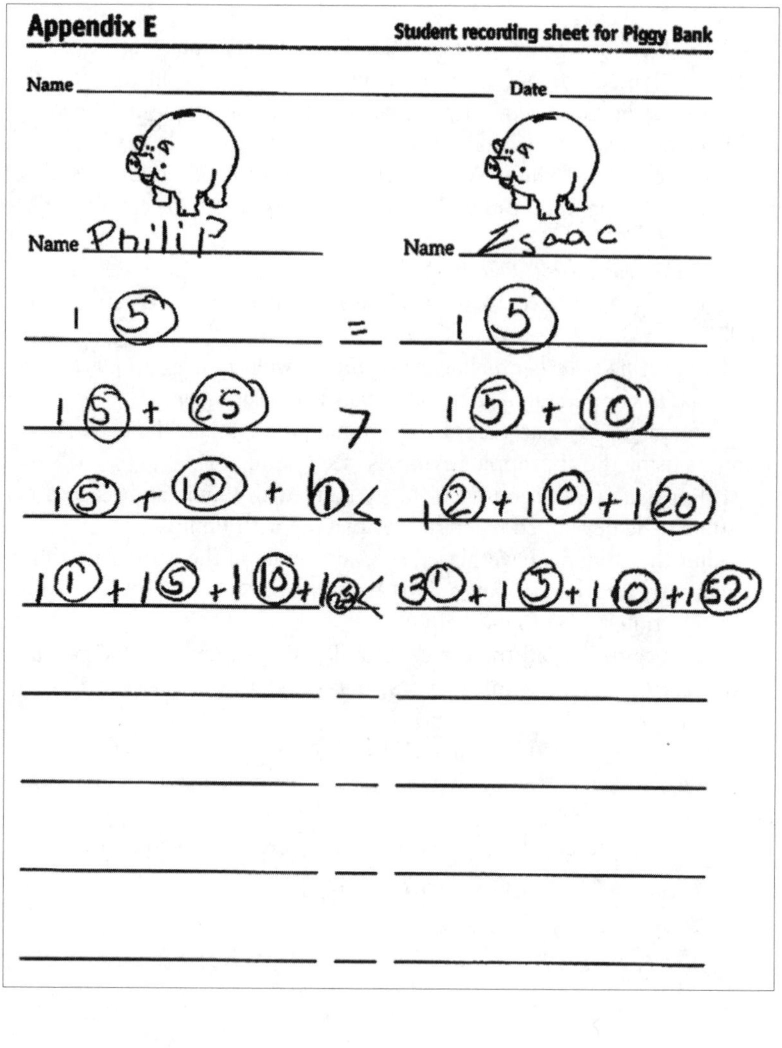
Appendix E — **Student recording sheet for Piggy Bank**

Name ______ Date ______

Name Philip — Name Isaac

1 (5) = 1 (5)

1 (5) + (25) > 1 (5) + (10)

1(5) + (10) + (1) < 1(2) + 1(10) + 1(20)

1(1) + 1(5) + 1(10) + 1(25) < 3(1) + 1(5) + 1(10) + 1(52)

lence. Although mathematicians eliminate extraneous equivalent expressions, this is difficult for children to understand. They often are willing to cross numbers out or to say that equivalent amounts don't matter, but they want to hold the amount "in storage" rather than eliminate it. Cynthia records "Isaac's cross-out rule" on the right under the word trade as the next step in the representation.

At this point most of the work children have been doing is within the context of money. Cynthia wants to generalize these procedures to the realm of pure number. Therefore, the next day she begins math workshop with a ten-minute minilesson consisting of a series of related problems, or strings. The focus is on deciding whether number statements are true.

"Let's warm up for math workshop today with a string of equations. And remember, let's look for shortcuts, like mathematicians! Here's the first one." She writes $5 + 4 + 10 = 10 + 5 + 5$ and asks, "True or not true?"

"Not true," Jasmine replies immediately. "It should say not equal."

Cynthia records the statement on the board and represents it on a series of cubes held to the chalkboard with small magnets (see Figure 5.5). The cubes are a concrete bridge for children who still need to see a spatial representation of the quantities.

Cynthia moves on to the next equation in the string. "How about this one?" She writes $9 + 5 = 8 + 6$. "True or not true?"

"True, because 9 is 1 more than 8, and 6 is 1 more than 5," Mia says with conviction.

Olivia is puzzled. "I don't get it."

Isaac explains, "It's like taking 1 from the 6 and giving it to the 8. You get $9 + 5 = 9 + 5$."

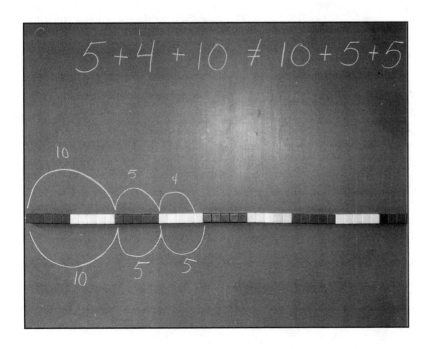

FIGURE 5.5
Representing Two
Expressions on a Bar
of Cubes to Examine
Equivalence

"Oh, cool—you could also do 10 + 4 = 4 + 10."

As Cynthia progresses through the string she begins using a double open number line in place of the cubes. (In an open number line only the numbers the children use are represented.) See Figure 5.6.

For the next problem, 13 + 8 + 6 = 5 + 9 + 13, children note that 13 has just been added to both sides, thus equivalence is maintained.

Then Cynthia challenges with 13 + 8 = 5 + 9 + 13 − 6 to introduce undoing. "True, or not true?"

"Equal. Because 13 is still on both sides, we can cross those out. Use Isaac's cross-out rule," Olivia suggests. "5 + 9 is 14 and 14 − 6 is 8. And 8 = 8." She still needs to do all of the arithmetic to feel certain of the equivalence.

Philip shares an alternative, more algebraic, strategy, employing the use of the commutative property in his justification. "My way is different. I started with the 9 and took 6 away. That left me with 13 + 8 = 8 + 13. And I know that is equal because the numbers can be turned around."

Cynthia challenges them to think about the cancellation as subtracting 6. "I'll share my way, but I don't know if it will work all of the time. Tell me what you think. When I wrote the problem, I used the problem before it: 13 + 8 + 6 = 5 + 9 + 13. I took six away from both sides, and since I knew the first problem was true I thought this one had to be, too. What do you think?"

"That works!" Jasmine exclaims. "All the time. Because if you add or subtract on both sides, and you use the same number, it stays the same."

Having gotten her young mathematicians to generalize, Cynthia represents symbolically the idea they are discussing. "So I'll use an *n* to represent any number like Jasmine said. If I write *n* + 13 + 8 + 6 = 5 + 9 + 13 + *n*, is this a true statement?"

"Yep." Olivia is confident. "It's like a mirror on the number line, like symmetry. What happens on the top happens on the bottom. And if you add a number and then take it away, it makes a jump but then you jump back to where you started. Looks like a banana!"

"Or like chalk dust!" Jasmine exclaims. "You make a mark and then you erase it. A number minus the same number is zero."

As children begin to understand that adding a number n to both sides of an equation does not disturb equivalence, they often overgeneralize and assume that if n is on only one side of the equation the equivalence is lost. Cynthia wants them to real-

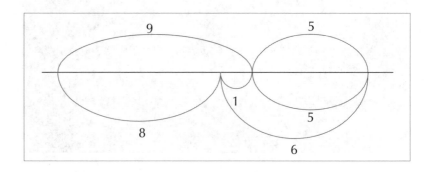

FIGURE 5.6
Representing 9 + 5 = 8 + 6 on the Double Open Number Line

ize that with some problems equivalence can be maintained if and only if n = 0 and to realize the importance of considering this possibility.

"So what about this one?" she says with a smile. "True or not true? I'll use *n* again to mean any number." She writes $8 + 6 = 5 + 9 + n$.

The first child to respond, Mia, overgeneralizes. "Not equal, because if you add a number on only one side it won't work," she declares emphatically.

Cynthia prompts discussion. "Turn to the person next to you and talk about what Mia just said. Do you agree?" After a few moments, Cynthia calls on Roxanna, who has been discussing the possibility of $n = 0$ with her partner, Haille. "Roxanna, what did you and Haille decide?"

"Mostly we agree, but what if the secret number is 0? Her idea doesn't work then."

Cynthia paraphrases. "So this is true only when *n* equals 0?"

WHAT IS REVEALED

Cynthia's second graders are developing a strong sense of equivalence and clearly have constructed a system that includes several rules of deduction that are regarded as valid in their community. For example, they have constructed compensation (what is added must be removed to maintain identity), commutativity, associativity, ignoring (canceling out) like amounts, and adding and/or subtracting *n* to both sides to simplify for analysis.

However, having a strong sense of equivalence and a variety of deductive rules to prove equivalence does not necessarily mean children will substitute equivalent expressions when they are computing. Encouraging them to do this is important, because a repertoire of strategies leads to better and faster mental arithmetic. For example, subtracting $321 - 189$, one might use the equivalent expression $332 - 200$. Adding 11 to each number gets the subtrahend to the landmark number of 200, making the problem easier to do mentally. One could also subtract 200 and add 11 back afterward. Both of these strategies—constant difference, and removing and adjusting—are much easier to do mentally than performing the standard regrouping algorithm using paper and pencil.

BACK TO THE CLASSROOM

Because the children are beginning to develop emerging ideas of proof, Cynthia wants to provide a more open-ended investigation. She begins by reminding the children of the story about the Masloppy family. The main character, Nicholas, loves to organize things. It was Nicholas who thought to organize the money in the piggy bank into rolls, and then into three equivalent piles. Cynthia tells her second graders that she once knew a boy just like Nicholas. He loved to organize things, too. One day he told her that he knew a fast way to subtract five. He said that instead of taking five away he added five and then took ten away.

She encourages the children to try the boy's strategy, and together they subtract 32 – 5. They get the same answer of 27 when they compute 32 + 5 – 10. It works! Then they set off, in pairs, to figure out whether the strategy will always work and why. After they've investigated for a while, Cynthia convenes a math congress.

"So Colleen and Juanita, you showed us several examples, and then you said, 'So we know it always works.' Have they convinced us that the strategy will always work?" *Cynthia prompts the community to consider what is necessary to make a convincing argument—what is sufficient to support a generalization. Her question goes right to the heart of algebraic thinking.* "Is just trying it out several times enough for us to be certain that it will always work? Maybe in the next problem you try, it won't work. We might not want to use this strategy if we can't be certain. How can we be certain?"

"I don't think trying it out lots of times is enough," Olivia exclaims. "I think we have to know why it works. And that's what's bothering me. I don't know why it works!"

"Hmm. Knowing why might help us be sure it will always work," agrees Cynthia. "What is going on here? Rosie, you and Jasmine have an interesting way for us to look at this. Show us what you did."

"We made a number line," Rosie says as she shows their drawing (see Figure 5.7). "We did the boy's strategy on top, and we did 32 minus 5 on the bottom. We got 27 both times."

"Let me write down an equation for what you said so we can all think about this. On the top of the number line is 32 plus 5 minus 10, and on the bottom is 32 minus 5." *Cynthia paraphrases to give everyone a chance to think about what Rosie has said. She writes $32 + 5 – 10 = 32 – 5$.*

Jasmine is excited. "It works because five and five make ten. That's it! I get it!"

"Who thinks you understand what Jasmine means? Who can put Jasmine and Rosie's idea in your own words?" Cynthia asks. *To get an inclusive discussion going, Cynthia needs to make sure that the ideas being offered are understood. Often teachers just say, "Does everybody get that?" and everyone nods or choruses yes. By asking, "Who can put Jasmine and Rosie's idea in your own words?" only children who can paraphrase the idea will put their hand up, and Cynthia can quickly tell who needs further clarification.* Several hands go up and Cynthia asks Ian to give it a try.

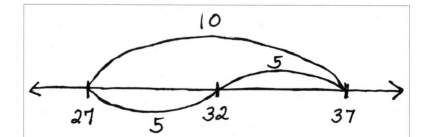

FIGURE 5.7
Rosie and Jasmine's Number Line Representation

"I think you mean that it works because the five and five make ten?" Ian looks quizzically at Jasmine who nods. "But if we did five and six it wouldn't work."

"Yeah. See?" Rosie draws another line. "The six is too big. It has to be five and five."

Cynthia reframes the comment. "So you are saying that if we want to take away six, we can't add five and take away ten. What would we add if we took away ten? Everybody, turn to the person next to you and talk about this." After a few minutes, Cynthia returns to the whole-group discussion. "Keshawn, what did you and Isaac decide?"

"We decided it would only work if we added four."

Isaac offers a new idea. "Or we could take away eleven and add five in order to subtract six, too."

Olivia joins the conversation. "Oh, yeah, wow, because five and six make eleven, and four and six make ten. The numbers have to add up to the number you take away."

The conversation continues as Cynthia challenges the children to consider other numbers. *Questions like this guide children to generalize and to examine the part/whole relations involved—getting right to the core of structuring. And as the part/whole relations are examined, equivalence is once again the focus.*

SUMMING UP

Understanding the equal sign, =, as equivalence and *not* as a symbol denoting "the answer is coming" is essential if children are to use algebraic expressions with meaning. The games and investigations in this chapter were set in contexts that required second graders to use equivalence and to examine equations relationally instead of only performing computations. They developed several "rules" that can be used in later years in chains of deductive reasoning as they work more formally to prove their thinking. These rules include compensation (what is added must be removed to maintain identity), commutativity, associativity, ignoring (canceling out) like amounts, and adding and/or subtracting n to both sides to simplify for analysis. For this they construct a second big idea: They learn to treat expressions as objects that can be operated on with these rules.

Children's work with equivalence is also tied to developing flexibility in using computational strategies such as compensation ($291 + 348 = 300 + 339$) or constant difference ($321 - 189 = 332 - 200$). As Einstein once said, "Everything should be made as simple as possible, but not simpler." The goal of algebra instruction should not be to dumb down the problems by making them simpler. Instead, the goal should be to support young learners to find their own ways to make problems as simple as possible. By inviting learners to find ways to simplify, rather than simplifying for them, we are inviting them into the world of the mathematician. We are involving them in the process of structuring problems. At the heart of structuring is simplification; it is precisely that process that brings beauty to the mathematician's creation.

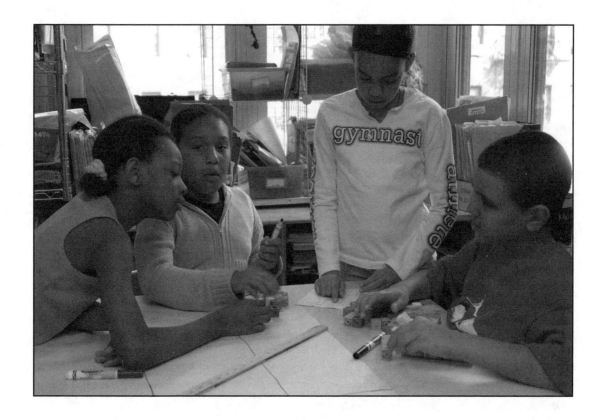

6 | VARIATION VERSUS VARIABLES

In symbols one observes an advantage in discovery which is greatest when they express the exact nature of a thing briefly and, as it were, picture it; then indeed the labor of thought is wonderfully diminished.

—*Gottfried Wilhem Leibniz*

But when the guidelines we give teachers are transmuted into lists of state-mandated jargon that are given an iconic status in the classroom, I don't think we're saving time for good instruction. I think we're stealing time from anything that actually contributes to a child's education.

—*Jonathan Kozol*

So far in this book we have seen children actively structuring the number system. They have considered even and odd numbers, "threeven" numbers, factors and multiples, equivalence, and more. Yet we have not begun to discuss what in the *public* eye is algebra—what about variables, what about x and y?

A quick glance at the walls of middle and high school classrooms reveals the presence of the rules of algebra. Posters proclaim the rules children need for success: order of operations; the binomial theorem; the quadratic formula; the slope intercept equation of a line. Often these posters cite the state standard their mastery represents. In California, a list of twenty-five standards nails down everything one needs to know for Algebra I, the standard eighth-grade course. Also in California, seventh-grade algebra standards make up a significant chunk of the high school exit examination, and failure to pass this exam has prevented tens of thousands of students from receiving a diploma in spite of successful completion of every other graduation requirement. The stakes are high!

Many researchers and authors have defined early algebra in ways that capture aspects of the activity we collectively recognize as algebra. But as we said at the outset, our objective is not to define algebra but to look at the *emergence* of algebra, in particular to identify the critical big ideas and strategies young children construct that are important landmarks for teachers to notice, develop, encourage, and celebrate. These developmental landmarks are *not* found on posters summarizing how to calculate or how to meet a particular state standard, but they are crucial for students to succeed.

Precisely because these landmarks are cognitive developmental leaps that often require the reorganization of previous ideas, they cannot be forced by segmenting material into predigested bits requiring sequential mastery. Too often, however, curricula and standards are organized in such bits: First it's solving one-step linear equations; then it's solving two-step equations; and when they are mastered we're on to three-step equations! Teachers are expected to follow sequential instruction, and some districts purchase standardized tests that monitor which students have mastered each step in the instructional sequence.

Some time ago a group of middle school teachers asked Bill for strategies on how to help students learn to solve three-step equations once they had "really mastered" two-step equations. Hoping to get the group to focus on student thinking rather than teacher actions, he began, "I'm not sure what you mean by a two-step or a three-step equation, but give me some samples and we can start by talking about what your students are doing." This cost him his credibility—how could a University of California mathematics professor not know the difference between a two-step and a three-step equation![1]

In many school districts the jargon has attained "iconic status" (to use Kozol's term), and the process by which students are expected to proceed from the two-step skill to the three-step skill is laid out in their texts, the district's standards, and the state's expectations. But there is much, much more to the development of the ability to solve equations than moving up step by step.

A BRIEF HISTORY OF THE DEVELOPMENT OF ALGEBRA

Historically, the first important step toward symbolic reasoning—as opposed to mere symbolic representation—occurred in the context of problem solving. Numerous ancient texts, including some from Babylonia and China, include problems in which information is given about some unknown quantity and readers are asked to determine its value. For example, a standard recipe for problems in Babylonian tablets begins, "I found a stone but did not weigh it." After some additional information, for example—"when I added a second stone of half the weight, the total weight was 15 gin"—the student is required to calculate the weight of the original stone (Stewart 2008).

The word *algebra* comes from the Arabic *al-jabr*, a term used by Muhammad ibn Musa al-Khwarizmi in 820 A.D.—the same mathematician who brought us the standard place value algorithms for arithmetic, such as long division. His seminal work *The Compendius Book on Calculation by Al-*

---

[1]To see how firmly this terminology is embedded in school algebra culture, search the Internet for *one-step linear equation* and *two-step linear equation*. You will find ample sources giving the definitions as well as detailed prescriptions on how to solve them.

jabr w'Al-muqabala laid out methods of calculation for solving six types of equations (Stewart 2008). *Al-jabr* means adding equal amounts to both sides of an equation; *al-muqabala* means subtracting equal amounts from both sides of an equation. Although these methods are recognizably similar to those taught today in a traditional algebra course, the latter term also has a general, and perhaps even more important, meaning—comparison. Comparing algebraic expressions gets to the heart of algebra—analysis occurs, relationships are determined, expressions are treated as objects, and equivalence is examined. Sadly, in the traditional teaching of algebra, emphasis is most often placed solely on the procedures to be used; the importance of the activity of comparison is overlooked.

As in the early Babylonian tablets, words were used by Al-Khwarizmi, not symbols. It took hundreds of years for today's algebraic symbolism to develop. Symbols first appeared for operations in the fifteenth century; the letters *p* and *m* were used as abbreviations for plus and minus. Soon after, the symbols + and − arose in commerce, where they were used by German merchants to distinguish overweight and underweight items. Mathematicians quickly began to employ them too. In 1557, the mathematician Robert Recorde invented the symbol = for equality. In his book *The Whetstone of Witte*, he wrote that he could think of no two things that were more alike than two parallel lines of the same length (Stewart 2008). (He used much longer lines than we do today, more like two open number lines, one on top of the other.)

Although some spotty use of symbols early on to represent unknowns has been found (for example, in Diophantus of Alexandria's *Arithmetica*, written around 250 A.D.), the move to symbolic notation for unknowns didn't gain momentum until the Renaissance. François Vieta was one of the first mathematicians to state his mathematical results in symbolic form. He used the consonants (B, C, D, etc.) to represent known values and used the vowels (a, e, i, o, and u) to represent unknown quantities (Stewart 2008).

However, it is not Vieta's use of letters that is important in the development of algebra, particularly since we use them differently today. What is important is that Vieta also made a crucial distinction between what he called the "logic of species" and the "logic of numbers." He argued that algebraic expressions could represent an entire class (species) of arithmetical expressions and that algebra should be seen as a method for operating on general forms. In contrast, arithmetic is a method for operating on specific numbers.

This distinction was a huge cognitive shift—a big idea. Before Vieta, equations were simply numerical relationships that allowed specific numbers to be substituted for the symbols x and y. Mathematicians were now challenged to distinguish algebra from arithmetic by treating algebraic expressions (such as $2x + 3y$) as mathematical objects that could be added, subtracted, multiplied, and divided without ever considering them as representations of specific numbers. Equivalent expressions could be exchanged. Algebra took on a life of its own, free from arithmetical interpretation (Stewart 2008).

This historical progression from solving problems with unknowns, to representing equivalence and comparing expressions, to the treatment of expressions as objects was a long one in the development of algebra. As will become evident in this chapter, many of the ideas that were cognitive leaps for mathematicians are big ideas for children, too.

IN THE CLASSROOM[2]

The fifth graders who have been working on frog-jumping problems (see Chapter 4) are moving on to solving linear equations, still within the context of frog jumping. Bill, their teacher, mentions Mark Twain's book *The Jumping Frog of Calaveras County* and explains that every year frog-jumping contests are still held at the Calaveras County, California, fairgrounds. He then presents this problem:

> MT is a bullfrog. He is world famous for his long jump. Every time he jumps he travels exactly the same distance. When he takes 4 jumps and 8 frog steps, it is the same as taking 52 frog steps.
> 1. How many frog steps are in 2 jumps and 4 frog steps?
> 2. How many frog steps are in each of MT's jumps?

Before the students begin working, they discuss some ground rules. What assumptions are necessary? In the story, frogs don't jump once and stop. Instead, they jump a number of times and then walk a little bit. If the various frog jumps are different lengths, possibly in multiple directions, it would be impossible to determine how long the jumps are from the information given. So for scoring purposes, it is assumed that the jumps are in a straight line and that their length will be determined according to this rule: *Whenever a frog jumps in an event, if the frog takes more than one jump, all jumps are equal in length. All frog jumps are measured in steps, and all steps are equal in length.*

Bryan and José start with the expression $4j + 8$ and interpret the fact that there are four jumps and eight steps in the sequence as meaning that each jump is two steps. They write (see Figure 6.1a), "MT did four jumps which is equal to 8 steps." They are not treating the expression $4j + 8$ as an object with j an unknown value—for them j must equal 2. Interestingly, they notice that $2j + 4$ must be half the distance of $4j + 8$ and determine its value as 26. However, most students solve the problem using direct calculations, beginning by undoing the operations that led to 52:

$$52 - 8 = 44$$
$$44 \div 4 = 11$$
$$22 + 4 = 26 \text{ frog steps}$$

[2]Portions of the dialogue of these fifth graders were previously published in an article by Fosnot and Jacob (2009).

The jump is treated as an unknown and arithmetic is used to determine the value. The value is then plugged into the next expression (2j + 4) to determine the numeric value—26. This strategy is reminiscent of the early algebra strategies found on Babylonian tablets.

Traditionally, teachers have often taught learners to use such procedures, but they employ the logic of the al-muqabala strategy (subtracting equal amounts from both sides of an equation). Learners are told, "To solve for j in the equation 4j + 8 = 52 get all of the numbers on the same side of the equation to simplify it. To do this, subtract 8 on both sides of the equation, because whatever you do to one side of the equation, you must do to the other. This produces 4j = 44, so divide both sides of the equation by 4 to get a result of j = 11." In contrast, in this frog-jumping situation, students have been invited to mathematize the situation in their own ways, and procedures to solve the problem have not been discussed.

A few students take other approaches. For example, Alyssa and Sam have constructed the following table:

4	8	52
2	4	26
1	2	13

Noticing that 2 is half of 4, and 4 is half of 8, they deduce that there are 26 frog steps in 2 jumps and 4 frog steps, since 26 is half of 52. They also draw a diagram of what the 4 jumps and 8 steps might look like (see Figure 6.1b).

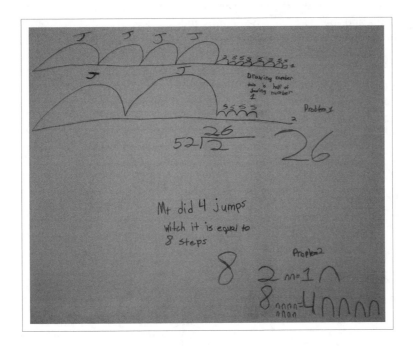

FIGURE 6.1a
*Bryan and
Jose's Work*

One purpose of this problem is to set up a situation in which the context of frogs jumping on a track can be developed and student strategies for determining unknown jump lengths can be discussed and represented on an open number line—a powerful model for comparing expressions. The numbers in the problem have been carefully chosen to encourage children to examine the relationship that Alyssa and Sam have used.

Bill invites Alyssa and Sam to share their diagram.

Alyssa explains, "Well, these are the jumps. Four of them. And these are the steps. Eight of them." She points to the related parts of their diagram.

"In your table, you wrote 2, 4, 26 underneath 4, 8, 52. What does that mean?" Bill asks. "Can you explain that using your diagram?"

This time Sam responds. "It's half of everything. That's why we did it. It really means that two jumps and four steps are twenty-six."

When several classmates look puzzled, Alyssa says, "See, here we have two jumps and four steps." She points to the lower number line in Figure 6.1b. "It's half of the first."

A few students now begin to voice their confusion. "How is it half? I don't see that."

"Well, two of these jumps are here, that's four, and four of these steps are here, and that's eight. So we doubled it."

Sam adds, "If you put two of the bottom diagrams together you get the top one."

Sam has brought up an equivalent relationship, and Bill wants to maximize this teaching moment. He challenges the students to consider both equations on the same number line, thereby emphasizing the equivalence. "So if you put two jumps and four steps and two jumps and four steps together you get the top diagram? Will they look exactly the same?"

For the next few minutes the class discusses the fact that if one considers only the lengths involved, it doesn't matter the order in which jumps and steps are drawn as long as there are the same number of jumps in each representation and also the same number of steps in each. Eventually they agree that if they halved the diagram again, it would show that one jump and two steps are equal to 13.

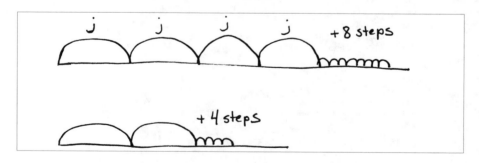

FIGURE 6.1b *Alyssa and Sam's Drawing*

At this point, Bill challenges, "Now that we've figured out how these diagrams are related, I'm wondering about a few more things. Can we use what we know to figure out three jumps and six steps? Talk with the person sitting next to you."

After a moment or two, Bill resumes whole-group conversation. "Maria?"

"Rosie and I think we can. It's 39, because a jump plus two steps is 13, so three times it."

Bill draws $j + 2$ three times (see Figure 6.2) and marks 13, 26, and 39 on the number line.

Juan and Tanisha offer a different strategy. "We did it a different way," Juan exclaims. "Tanisha and I subtracted a jump and two steps. That makes 3 jumps and 6 steps, too."

"Is this what you mean?" Bill draws four jumps and eight steps and then crosses out one jump and two steps. "How do you know how many steps that is?"

"We subtracted. Fifty-two minus thirteen equals thirty-nine."

Treating algebraic expressions as objects rather than simply as procedures is an important landmark on the landscape. To ensure that the students examine this idea, Bill prompts, "Turn to the person you are next to and talk about what Juan and Tanisha did. Why are they subtracting?"

Alyssa shares her excitement. "Oh, that's really cool. It works. And that means we could make lots and lots of ways. We can half or double, and we can add or subtract."

Juan adds another critical concept—the generalization of the idea of equivalent algebraic expressions. "Actually we can multiply or divide by any number, not just double or half. It's like if something is equal to something, then you can use those things and add 'em to other equal things."

WHAT IS REVEALED

Interestingly, the fact that the jump was a length of 11 steps was not part of the children's discussion. The class already knew this (most students determined this using an undoing procedure). Instead the conversation focused on the relationship between diagrams representing $4j + 8$, $2j + 4$, and $j + 2$. Based on Alyssa and Sam's representations, the students understand that $2 \times (2j + 4) = 4j + 8$ and that $4 \times (j + 2) = 4j + 8$, although these expressions were not written down in that form.

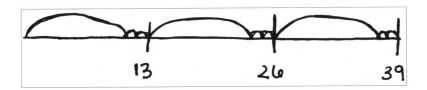

13   26   39

FIGURE 6.2
j + 2 three times

In due time, the students will use these symbols. For now, it is significant that they are manipulating chunks of diagrams in meaningful ways—in effect treating the expression $2j + 4$ as an object. The arc representing j is not merely an unknown number—it is a meaningful quantity that can be manipulated, and in this context it has a visual representation.

A new type of equivalence has arisen in this conversation—students are seeing that *algebraic expressions may appear different yet can be equivalent.* That four jumps and eight steps is the same as repeating two jumps and four steps twice is true independent of the length of the jump. The variable j used this way represents true variation. In Vieta's words, it represents the logic of species rather than the logic of numbers. Instead of viewing $2 \times (2j + 4) = 4j + 8$ as being the result of a symbolic rule, the variable has a representation as an object—an arc that can vary—and the steps can be understood as lengths too. These ideas—that an expression can be thought of as an object and can appear different yet be equivalent, and that these relationships hold for various lengths—are big ideas that will become further elaborated as students have more opportunities to investigate equivalence and representations on the number line.

BACK TO THE CLASSROOM

Now that his students have multiple ways to represent equivalent jumps and steps on a double open number line (using the top and bottom of the line), Bill invites them to explore a new series of problems, the two-trial jumping contest, in which three frogs—Cal, Sunny, and Legs—each complete two jump/step sequences that land them in the same place:

- When Cal jumps 3 times and then takes 6 steps *forward*, he lands in the same place as when he jumps 4 times and then takes 2 steps *backward*.
- When Sunny jumps 4 times and then takes 11 steps *forward* he lands in the same place as when he jumps 5 times and then takes 4 steps *forward*.
- When Legs jumps 2 times and then takes 13 steps *forward* he lands in the same place as when he jumps 4 times and then takes 5 steps *backward*.

Bill reminds everyone of the frog-jumping rule—*whenever a frog jumps in an event, if the frog takes more than one jump, all jumps are equal in length; all frog jumps are measured in steps, and all steps are equal in length*—and students set off to determine which frog has the biggest jump and therefore wins the contest.

At first many students have trouble comparing jumping scenarios (3 jumps + 6 = 4 jumps − 2) because they create two open number lines and either don't start jumps at the same point or don't align the jumps equally. Consequently, they either have to redraw their work or draw various curves to indicate which parts of one diagram correspond to parts of the other. (See Figures 6.3a, b, and c.)

Problems like these can be solved by guess-and-check methods, and when students use this approach they have made sense of the context and the question. But it does not lead to the type of structuring and sense making about expressions with variables that Bill hopes to generate, so he encourages students to represent the problem using the relationships provided. He confers with Thomas and Alyssa, who are working on Sunny's jumping data.

Thomas offers a guess. "What if the jump is 10? Let's try that—4 times 10 is 40, plus 11 is 51."

"Okay. I'll do the other one for Sunny." Alyssa plugs 10 into $5j + 4$. He has to land at the same place, right? Five times 10 is 50, plus 4 is 54. Doesn't work."

Bill suggests, "Sometimes when mathematicians feel stuck they begin by modeling the problem. Have you thought of drawing the jumps and the steps, like on a track, or a number line?"

Thomas and Alyssa look surprised at Bill's suggestion. "How can we do that on a number line? We don't know what the size of the jump is!"

"Can you draw the jumps on a track?" *Bill offers the same suggestion but keeps it grounded in the context.*

"I guess." Thomas draws a track with 4 jumps and 11 little steps. The context helps him realize what he is doing.

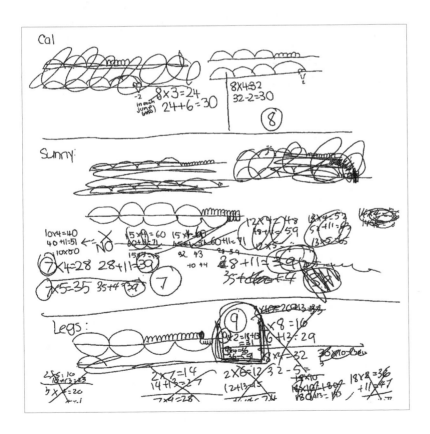

FIGURE 6.3a
Poster for the
Jumping Contest

"I'll do the other one," Alyssa offers. She draws a track with 5 jumps and 4 steps, but her tracks and jumps are different sizes than Thomas' are.

Bill attempts to focus them on the equivalence. "It looks like your jumps and your track are bigger than the ones Thomas made. Does that matter?"

Thomas exclaims, "Yeah. It's the same track, and they have to land at the same spot! You have to make yours exactly like mine."

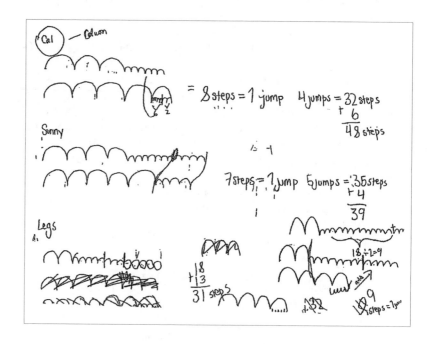

FIGURE 6.3b
Poster for the Jumping Contest (continued)

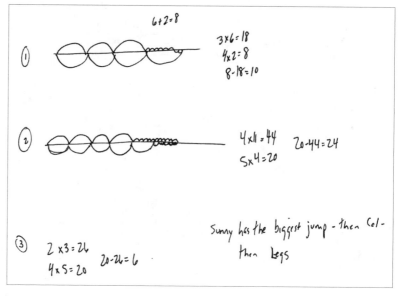

FIGURE 6.3c
Poster for the Jumping Contest (continued)

By coordinating their representations, Thomas and Alyssa will come to real-ize how the jumps and steps are related. At first students struggle to find common meeting points on the line, but this struggle is an important precursor to develop-ing the big idea of equivalence and deriving later cancellation strategies.

Bill now suggests they use just one line. "Since it's the same track, maybe you should just draw one line. You could put Sunny's first trial of 4 jumps and 11 steps on the top and his second trial of 5 jumps and 4 steps on the bottom. And you're right, Thomas—they have to land at the same place."

Thomas and Alyssa start drawing on the same track, each doing one side, but the jump sizes are still not the same. Bill helps them stay grounded in the context. "I thought you said the frog's jump sizes were the same?"

"Oh, yeah—the frog-jumping rule. Let's start over."

When they begin using double number lines, many students find that after an initial attempt to represent the problem, they have to start over. Bill encour-ages, "It's fine to start over when you are making double number lines. After all, you don't know how big the jump is when you start, so it makes sense that you may have to redraw."

The next day, after students have had time to create posters of their work, Bill convenes a math congress and picks Maria, Yolanda, and Sam to present their work, each of them dealing with one part of the problem.

Maria begins discussing her work on Cal. (Her work is shown in Fig-ure 6.4.) "I drew three jumps and six small steps, and another three, I mean, four jumps take away two small steps. And I drew this a certain way showing the line because I decided to ignore these three jumps so I could see how the two parts—six steps and the jump taking away the two steps—were equal. So. . . ." Maria hesitates.

After a moment Bill asks students to tell a partner in their own words what Maria has done so far. "Turn and talk to the person next to you. Try to put Maria's strategy in your own words. Talk about what the big bar in her drawing is."

After a few minutes, Bill asks Alberto to share his explanation. "I think it is kind of like a separator, because the three jumps on the top and the

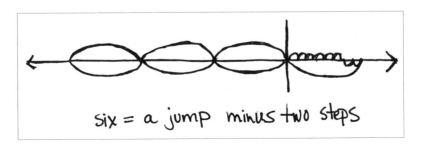

FIGURE 6.4
Maria's Work on Cal's Jumps

three jumps on the bottom both meet up. But we want to see how the six little steps and the big jump minus two steps meet up because it is supposed to end up in the same place—because it's not really important where the other three jumps are at. It's really just important to find out how they meet up."

Bill paraphrases to focus on what has been separated out. "You say this is like a separator. You think Maria separated the stuff that doesn't matter so much?"

Alberto clarifies, "Well, it is not that it doesn't matter. It's just that you could tell that they automatically meet up and that you don't have to worry about that part."

Knowing how difficult it has been for many of the students to represent the two equations on the same line, Bill continues probing. "You could tell they meet up because . . ."

Alberto says confidently, "Because three jumps and three jumps are the same thing!"

Maria now uses her hand to show on her line how the sum of 6 and 2 is equivalent to one jump, successfully convincing her peers that the distance of Cal's jump is 8 steps.

WHAT IS REVEALED

The Importance of Representation

The double number line representation is a powerful tool for helping students examine equivalence and variation. Most students find it easier to reason with the double number line than to guess and check. It is also more interesting for them to draw the diagrams than to guess with numbers. This is an important developmental transition. At the same time, working with the diagrams does not guarantee an answer on the first go-round (Thomas and Alyssa's struggle is common). When drawing two diagrams, children's endpoints often don't match and they are unsure how to proceed (see Figure 6.3b, page 102).

There is a lot at stake in setting up these three frog-jumping problems. In Cal's case, the students have to create a representation for $3j + 6 = 4j - 2$. This is a so-called three-step equation (although many students can solve it in two steps). Students who memorize the steps for solving such equations when they do not have a mental image of what the equality represents often confound the procedures as they work with symbols. They ask, "Do you add or subtract $3j$? Do you add 2, subtract 2, or what?"

As students struggle to represent three jumps followed by six steps forward in relation to four jumps followed by two steps backward, they are constructing the big idea that *you can work meaningfully with expressions containing unknown quantities, even if you don't know their value.* Freudenthal (1973) argued that concepts are the results of cognitive processes and not the other way around. Working with diagrams facilitates important cognitive processes. When students diagram the two jumps, they eventually line them up with the same beginning and end points. The context of the frogs jumping on tracks helps them realize the equivalence. As they study their equivalent representations, they notice that the first three jumps in each must line up. This is another big idea, that *when variables are used repeatedly in equivalent expressions they must represent the same thing* (in this case the unknown jump length). At this point the students are working meaningfully with unknown quantities, even though they don't know what they are. The jumps in the diagrams are representations of a variable and a mental object is forming alongside them.

The Importance of Redrawing

In order to get the two diagrams to align, students often have to redraw their initial try. They forget to line up beginning and end points, or they aren't careful to draw jump lengths the same (see Figure 6.3a, page 101). Sometimes it takes many tries. Revision is an important part of the learning process.

Redrawing is important because it raises issues about equivalence and leads to the notion of *variation*. The equivalence in *three jumps followed by six steps forward is the same as four jumps followed by two steps backward* expresses a relationship between jump sizes and step sizes. This is the essence of the big idea of variation, that *variables describe relationships*. Too often students believe a variable is simply an unknown number to be found, a habit of mind likely developed from solving equations procedurally. But variables represent more than unknowns, they represent (Vieta's word) a species. This notion of variation underlies the notion of *function*, probably the single most important abstract construction in higher mathematics. All of these big ideas are packed into making sense of three-step equations.

BACK TO THE CLASSROOM

As Bill's math congress continues, Yolanda presents her solution to Sunny's jump length, $4j + 11 = 5j + 4$ (see the diagram in Figure 6.5).

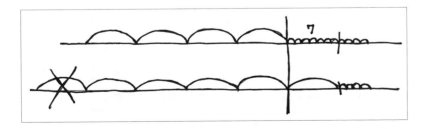

FIGURE 6.5
*Yolanda's Solution
to Sunny's Problem*

"Well," she begins confidently, "I knew that there were four equal jumps both times, so one jump and these four steps [*pointing to the four frog steps*] had to equal eleven steps. I saw four here [*comparing the 4 to the 11*] so this and this were equal [*pointing to one jump and four frog steps and 11 steps*] so that left seven frog steps here. That meant it has to equal one jump because there were only seven left and there were four before."

"I get it," Maria interrupts, smiling. "It's pretty much the same as what I did with Cal. You took away the jumps because they didn't really help."

"They both used a separation bar, too," adds Alberto. "I think that was the biggest similarity, because it was like the root of their, um, strategy. So I think that was the biggest . . ."

"So the separation bar enables you to take away things that are unnecessary?" Bill asks. *Are ideas about cancellation emerging?*

Several students nod affirmatively.

Maria offers a new image. "Like a storage box, kind of."

Bill probes their sense of variation. "Does it matter that we don't know the sizes of the jumps in the storage box?"

"No because they're equal—the size doesn't matter," Alberto says with conviction.

No one objects, and Bill asks Sam to share his analysis of the data on Legs's jumps, $2j + 13 = 4j - 5$. He presents what might be considered a generalization of the cancellation strategy (see Figure 6.6).

"Depending on the difference in the number of jumps and whether the steps are forward or backward you can always figure out how many steps are in a jump," Sam begins. "For Legs it is thirteen plus five divided by two is nine. That's how many it was, because there are two jumps and then thirteen steps [*draws*] and then there are four jumps and five steps backward, so since there are two extra jumps that's why you need to divide by two." He writes $13 + 5 = 18 \div 2 = 9$, which after a short discussion on the proper use of the equals sign he rewrites as $13 + 5 = 18$ with $18 \div 2 = 9$ below it.

Bill asks for clarification. "Show us on the picture—where is the thirteen plus five?"

Sam obliges. "If these two were the same jump it would just be thirteen plus five, because two jumps plus thirteen equals up to here, and then the five would go back to there because they have to land in the same place. So you need to know thirteen plus five to get here. If this was all one jump it

FIGURE 6.6
*Sam's Generalization
of the Cancellation
Strategy*

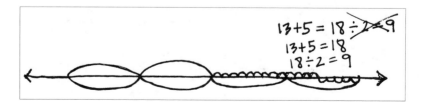

would just be thirteen plus five. But since it is two jumps you need to divide it by two because they are separate jumps."

Bill relates what Sam has described to the earlier discussion. "Do you think Sam used the storage box idea in his shortcut?"

"Yes," says Juanita, and many of her classmates nod in agreement. "It was kind of like an invisible line even though he didn't draw it, because the first two jumps on the top and the first two jumps on the bottom he didn't care about because they wouldn't help him solve the problem. He just paid attention to the two last jumps and the thirteen little steps, so technically he is using it but like an invisible line. He simplified the problem to be two jumps equals thirteen plus five, and then he divided each by two to get one jump."

"Do you agree, Sam?" *Bill wonders whether Sam sees that he has done the elimination mentally.*

"I guess . . . maybe . . . I don't know."

WHAT IS REVEALED

Cancellation is a process employed in symbolic reasoning that is typically taught when students are learning equation-solving rules. In the previous discussion, the class, working as a community, is devoting part of their efforts to constructing and defining an algebraic rule whereby one removes equal amounts of an unknown from both sides of an equality. The conversation centers around certain lines drawn to locate equivalent points on two open number lines. Some students have trouble figuring out where to draw lines on their diagrams. The symbolic issue is how to deduce $6 = j - 2$ from the earlier representation of $3j + 6 = 4j - 2$—that is, how to "cancel" $3j$ from each side of the equation.

The terminology used in this discussion gives some clues about the students' development of a cancellation rule. In explaining the bar Alberto first describes it as a *separator*. This terminology is appealing, because many students seem unwilling to remove the equal amounts from consideration. They view the separator as breaking the problem into two parts, even though they recognize the equivalent amounts on the left are not necessary to answer the question. This hypothesis is supported later when Maria refers to the separator as a "storage box." The equal amounts in the unknowns are not thrown away; they are stored, not "cancelled." The students also seemed to prefer the word *unnecessary* rather than the words Bill used—*ignore* and *remove*.

On the other hand, their terminology could result from the context. In the frog-jumping contest, jumps aren't thrown away or removed. Referees consider all jumps and all steps in determining the length of a single jump. Nevertheless, Sam does appear to cancel equal amounts and remove them from consideration. His rule considers only the extra jumps and the forward and backward steps, and on his written paper he uses equals signs to represent the result of operations rather than equivalence, writing $13 + 5 \div 2 = 9$, together with the check $9 \times 2 = 18 + 13 = 31$ and $4 \times 9 = 36 - 5 = 31$.

This misuse of equality is corrected during the congress, but only after Bill has uncovered the sequence of operations Sam is describing, which is a clue to Sam's thinking. Sam creates a diagram that includes all the jumps, indicating he understands the representation, but he does not include a separation line and his discussion includes only the right portion of the diagram. That the other students realize he is using a version of their earlier rule in the form of a "hidden line" indicates the community is making progress toward identifying cancellation as a deductive rule. But it is not clear that Sam sees that he is using that rule. In fact, on the back of his paper he records his justification to the second problem as: "4 jumps + 11 steps is the same as 5 jumps + 4 steps, so the difference between 11 steps and 4 steps is the difference between 4 jumps and 5 jumps." Assuming he used similar reasoning in the third situation, we can understand why he was not convinced he was using an invisible line. He is able to work with the remaining jumps (as if they were separated out), but he remains aware of the context.

These students will, in time, symbolize their work with conventional algebraic expressions and make sense of symbolic rules of reasoning, such as canceling equivalent amounts. But the conversations in this congress illustrate why it is crucial to allow students time to first make sense of the algebraic operations in context and to use representations (in this case double number lines) as a tool. As fifth graders they are conversing about how to make sense in solving three-step linear equations. Many big ideas are on the horizon! These same problems, set in a symbolic fashion without the proper grounding, can trip students up. There is time between fifth grade and twelfth grade for sense making in algebra. We need to find that time—and take that time.

SUMMING UP

Algebra has traditionally been taught as a collection of rules to solve particular types of problems. These rules do govern how symbolic expressions are manipulated and how equations involving them can be solved, but algebra instruction now focuses almost entirely on the formalism in which these rules are embedded. The jargon we use to describe the progression of processes has become iconic, and students no longer have the time they need to construct the big ideas that underlie the important mathematics— expressions as objects, equivalence, variation, the fact that expressions with variables describe relationships.

Representing expressions with variables on open number lines facilitates the construction of big ideas. Four frog jumps and eight frog steps can be visualized, even though one may not know how large the jumps are in relation to the steps. This visual image helps students work with expressions as objects. They can then, using double number lines, investigate equations involving expressions. Variables take on a special value governed by the equation (or equations). Often, because variables describe relationships (variation), these double number line representations have to be redrawn as part of the problem-solving process. But this redrawing process is a good thing. It leads to a deeper understanding of variation and the development of powerful strategies for solving for unknowns. Most important, these strategies and ideas are generated and owned by the students and thus employed with a robust understanding, empowering them as developing mathematicians.

7 | FURTHER HORIZONS
Integers and Equivalence

We are usually convinced more easily by reasons we have found ourselves than by those which have occurred to others.

—Blaise Pascal

In my view, the primordial and—in most cases for most people—the final goal of teaching and learning is mental objects. I particularly like this term because it can be extrapolated to a term that describes how these objects are handled, namely, by mental operations.

—Hans Freudenthal

Why is $-4 - -5 = +1$ true? Most students are not convinced when told, "It must be so." Nor are many adults. Parents and educators sometimes offer learners explanations to try to trigger understanding—for example, "When you get rid of your debt, you gain money," or, "Subtraction and addition are inverse operations and since subtraction takes away what was added, if we have $1 + -5 = -4$ we also have to have $-4 - -5 = +1$." But just as often, students are simply given rules (two minus signs make a plus) and asked to use them even if they don't understand them. The unfortunate result is that they implicitly accept mathematics as not needing to make sense.

But it is critical that young, developing mathematicians believe that mathematics makes sense. We don't yet have definitive or comprehensive research on the best way to teach operations with integers, but given the fluency with integers expected of students when they take algebra courses, it's important to provide some glimpses of contexts and models that help students investigate the big ideas involved in the addition and subtraction of integers, as well as some of the underlying theory.

The notion of number evolves progressively. It begins with using small subitizable units and counting, constructing cardinality, and building a measurement model for number, each of which represents significant cognitive leaps. Then, through investigating fair shares and part-to-parts relationships with fractions and decimals, the notion of number gets extended to include rational numbers. In each case context is a powerful learning tool, and the everyday actions of counting, measuring, and dividing up root the extension of ideas about number in familiar experiences.

But negative numbers are somehow different. We don't typically observe them as a quantity[1] either to count or measure, and we don't typically experience antimatter, which wouldn't necessarily provide a good model for them anyway. The negative numbers are mental objects having enough properties in common with the other numbers that we allow them into the family—into our structuring of the number system.

If we take seriously the idea of algebra as a structuring activity, then *extending counting numbers to include negative integers requires that we construct new mental objects.* There are no shortcuts. Current approaches typically either involve demonstrations on a number line (going backward to include negatives) or equivalences using + and − chips in which the chips cancel each other out—for example, three positive (red) chips added to five negative (yellow) chips equals two negative (yellow) chips. These approaches do capture the key features of integers and their arithmetic and, in fact, are linked to important applications. However, it is very easy for them to become rule driven and for students to arrive at answers based on their actions with the manipulatives rather than operating mentally.

The context of net change, which is built on equivalence, is instrumental in understanding integers. Net gain and net loss are both mental images—generalizations—but they arise from examining data and noticing patterns in realistic situations.

TEACHING AND LEARNING IN THE CLASSROOM[2]

Patricia Lent begins her second-grade math workshop by reminding her students about taking the subway to Central Park for a field trip:

> Remember when we took our birding trip to Central Park and we rode the subway? We all got on the same subway car. It was really crowded. None of us had seats. At the first stop, a few of us got seats. Then at the next stop, a few more of us got seats. The car kept getting emptier and emptier, until by the time we got to the park, everyone had a seat. In fact, we were practically the only people on the car. During our subway ride a lot of people got off the car, but people also got on the car at nearly every stop. I wondered how our subway car kept getting emptier if there were people getting on at every stop.
>
> I was still puzzling about this on my way home from school yesterday. I got on the local train at Chambers Street. When the

---

[1] While we do read temperature below 0, these numbers do not really have the features of negative numbers—they measure thermal energy, which is always positive.

[2] The dialogue and work samples of these second graders were previously published in an article by Lent, Wall, and Fosnot (2006).

train left Chambers Street, I counted ten people in my car. At Franklin Street, the next stop, some people got off, and some people got on. I couldn't see exactly how many people got off and on, but when the train left Franklin Street, there were fifteen people in the car. I'm wondering what could have happened. How many people could have gotten off? How many people could have gotten on?

Although most members of the class realize the problem has multiple solutions, a few students initially see it as having a single solution. And only a few of the students who realize the problem has multiple solutions seem to understand there might be a finite number of solutions. A few students use some kind of system to generate solutions; however, most find new solutions by random searching. They represent their solutions in a variety of ways—diagrams, open number lines (a common model they have used for arithmetic), prose, and equations (see Figures 7.1a, b, and c).

During the first math congress, Patricia and Cathy (who is also working in the classroom) focus on the different ways students have represented and organized their solutions. The class is particularly intrigued by Uriah and Mia's use of what they term "a pattern" (specifically, decreasing by one the number getting off and decreasing by one the number getting on) to generate eleven solutions beginning with $10 + (-10 + 15) = 15$, $10 + (-9 + 14) = 15$, and so on, ending with $10 + (0 + 5) = 15$. "If you use a pattern," one student says, "you'll get more ways." However, although many of the students seem to understand the landmark strategy of using a systematic procedure to generate a lot of solutions, the big idea that one might be able to use this strategy to generate *all* the solutions seems out of reach for most. When Chynna and Nyima share their system for systematically keeping track of

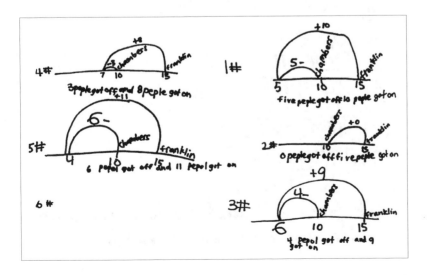

FIGURE 7.1a
Subway Problem Work

which solutions they have found, few of their classmates seem to understand why being systematic is important.

The next day, to provide further experiences with these ideas, Patricia and Cathy present a new but similar problem. They tell the children that ten people are in a subway car at Chambers Street but after the next stop there are seventeen people in the car. As the partners get to work, spirited

FIGURE 7.1b
Subway Problem Work (continued)

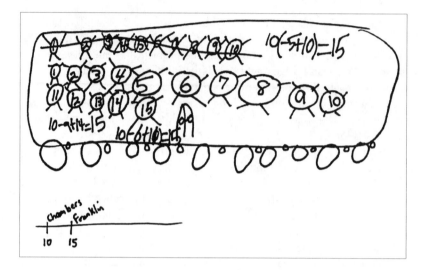

FIGURE 7.1c
Subway Problem Work (continued)

1. $10 + (-10 + 15) = 15$
2. $10 + (-9 + 14) = 15$
3. $10 + (-8 + 13) = 15$
4. $10 + (-7 + 12) = 15$
5. $10 + (-6 + 11) = 15$
6. $10 + (-5 + 10) = 15$
7. $10 + (-4 + 9) = 15$
8. $10 + (-3 + 8) = 15$
9. $10 + (-2 + 7) = 15$
10. $10 + (-1 + 6) = 15$
11. $10 + (0 + 5) = 15$

We found 11 Different ways to answer this problem. We made a pattern with the numbers.

conversations about patterns and possible solutions begin: "Let's do a pattern, so we can find lots of ways!" "How many ways do we have?" "How many do *they* [nearby pairs of classmates] have?"

Working systematically does help students generate additional solutions. However, using this strategy doesn't necessarily lead them to ask the critical mathematical question, "Do we have them all?" For instance, Ava and Danielle successfully use a table to find and record eight solutions beginning with "7 people got on, 0 got off." They stop at "14 people got on, 7 got off" not because they've found all the solutions but because they run out of room on their paper. "We found out that we made a pattern," they write on their poster, "and we are going to do it again." Devin and Philip also use this "pattern" to find five consecutive ways, but stop after five when they too run out of room.

Tova and Haille end their search for solutions for a different reason. In their investigation the day before, they used an open number line to represent their thinking, jumping forward to record the number of people getting on and then jumping backward for the people getting off. Focused on this strategy and forgetting about the context, they ended up with some impossible solutions (30 people getting on and 25 people getting off, even though there were only 10 people to begin with). This time they start at ten and jump backward first, then jump forward. "We figured out it was easier to minus first," Tova explains. They also pair each open number line with a diagram of the subway car "just to make sure." Their pattern begins with $10 - 5 + 12$ and continues until they reach $10 - 9 + 16$. They stop here, reasoning that since Patricia had to stay on the train, no more than nine people could get off.

Patricia challenges them to think about other possibilities. "Do you think you have them all?"

Haille replies, "Well, we can't think of any more, but Ava and Danielle have more than we do."

"Let's see, you started at −5 and went all the way down to −9," Patricia probes. "Could you extend your pattern in the other direction?"

"Oh, now I see!" Haille responds. Excitedly they tape a blank sheet of paper above $10 - 5 + 12$ and extend the pattern up to $10 - 1 + 8$ (see Figure 7.2). They stop here, since the story is "some people got off and some people got on" and thus $10 - 0 + 7$ is not a valid solution.

The math congress for this second investigation begins where the previous one ended—looking at different representations and discussing students' systematic approaches. First Ava and Danielle share their table; then Devin and Philip share their poster. Their classmates note that both partnerships have used "patterns," but at first they think the two posters show different solutions. Ava and Danielle began with people getting on, while Devin and Philip began with people getting off, and their numbers are therefore in a different order. Eventually, they realize that Devin and Phillip's "$10 - 2 + 9 = 17$: 2 people got off then 9 people got on" is the same as Ava and Danielle's "9 people got on, 2 got off." Mia, commenting on the rearrangement of symbolic representations, explains, "It's the same thing just put in a different way." Isaac points to the functional nature of the patterns and exclaims, "It's like a machine!" Interestingly, Isaac had no previous experience with input/output function machines that we are aware of, which makes his comment so remarkable. But he loved to draw intricate gigantic imaginary machines that did things only he understood and in the block area he often built structures that marbles and balls would roll through.

FIGURE 7.2
*Haille and
Tova's Work*

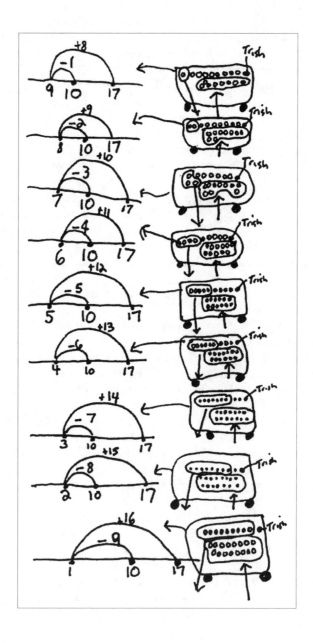

Given the context of people leaving and getting on, it can be argued that $-2 + 9 = 9 - 2$ is really a series of operations on 10, with whole numbers representing quantities of people ($10 - 2 + 9 = 10 + 9 - 2$), and doesn't require the construction of negative and positive integers. However, over the next few days the emergence of a mental image of *net change* begins to incorporate a truer sense of positive and negative quantities as children describe these results as equivalent to the overall end result of people getting on and off.

BACK TO THE CLASSROOM

The third investigation Patricia and Cathy initiate involves net loss. (The idea of net change had come up briefly in the second congress when Devin presented his poster and explained, "The more that get off the higher the number to get on." However, it had been overshadowed by the students' eagerness to compare representations and identify systematic approaches.) This time, the train continues on from Franklin Street with 17 people on board. At the next stop—Canal Street—some people get on, some people get off, and there are 15 people on board—fewer people, which is what happened on the class field trip, but the net change is small. Patricia and Cathy begin the discussion by looking at Austin and Isaac's series of equations (see Figure 7.3) and Ava and Danielle's table (see Figure 7.4).

Devin, generalizing beyond the specific instances he observes, says, "The number that minuses has to be two numbers bigger than the number that plusses." This idea spreads and is picked up by a number of the students.

Chiara: On every one, it minuses two.

Nyima: Two got off, zero got on, and that equals two. All the way down on both sides it's always two.

Sam: The numbers are two more here than the other side. Four is two more than two, seven is two more than five.

Aidan: It's three and one, four and two, five and three, six and four. It's like counting by two but using the numbers in between.

Delia: I understand that everything is two away—if four got off, two got on, because it's two away from the number you started with.

Michael: Each time two more people got off than got on.

In order to see net change in the posters, these second graders must shift their attention away from the pattern going down the columns and focus on the across relationship between the pair of numbers in each row. More to the point, they must shift from the numbers of people getting on and off to an abstraction—a mental image of the resulting negative result. Although many students are perplexed, they

*know they have a responsibility to articulate what they do understand and what
they need help with. These young mathematicians are working very hard.*

Austin: I have a question. I get that they go in order, and the minus is more
than the plus. So if you take away more, do you add less?

Simone: If you take away less you won't get to the right number.

Delia: Because fifteen is two less than seventeen.

Nyima: Because seventeen is two more than fifteen. If you're taking away
more, you're ending up with less.

Chynna: Seventeen minus two equals fifteen. So I think that relates to the
other question, why off is two more than on.

Mia: I don't fully understand. It has to do with the pattern—one getting
higher, the other getting lower.

Michael: What Chynna is trying to say, and what Nyima is trying to say, is
everything you do, it has to have two more people getting off than getting on.

WHAT IS REVEALED

At this point in the discussion the children are beginning to get a mental
image of net loss, and this mental image of −2 is seen as equivalent to many

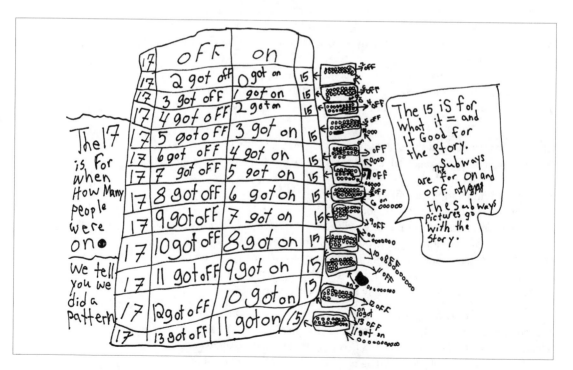

FIGURE 7.3   *Austin and Isaac's Work*

other expressions. For example, they understand that $+15 - 17$ is equivalent to $1 - 3$ and $2 - 4$. Although the net loss is still directly linked to the result of people getting on and off—operations with positive numbers—the result itself is an abstraction.

Operating with integers is not yet on the horizon for these second graders—this topic does not usually appear in the curriculum until middle

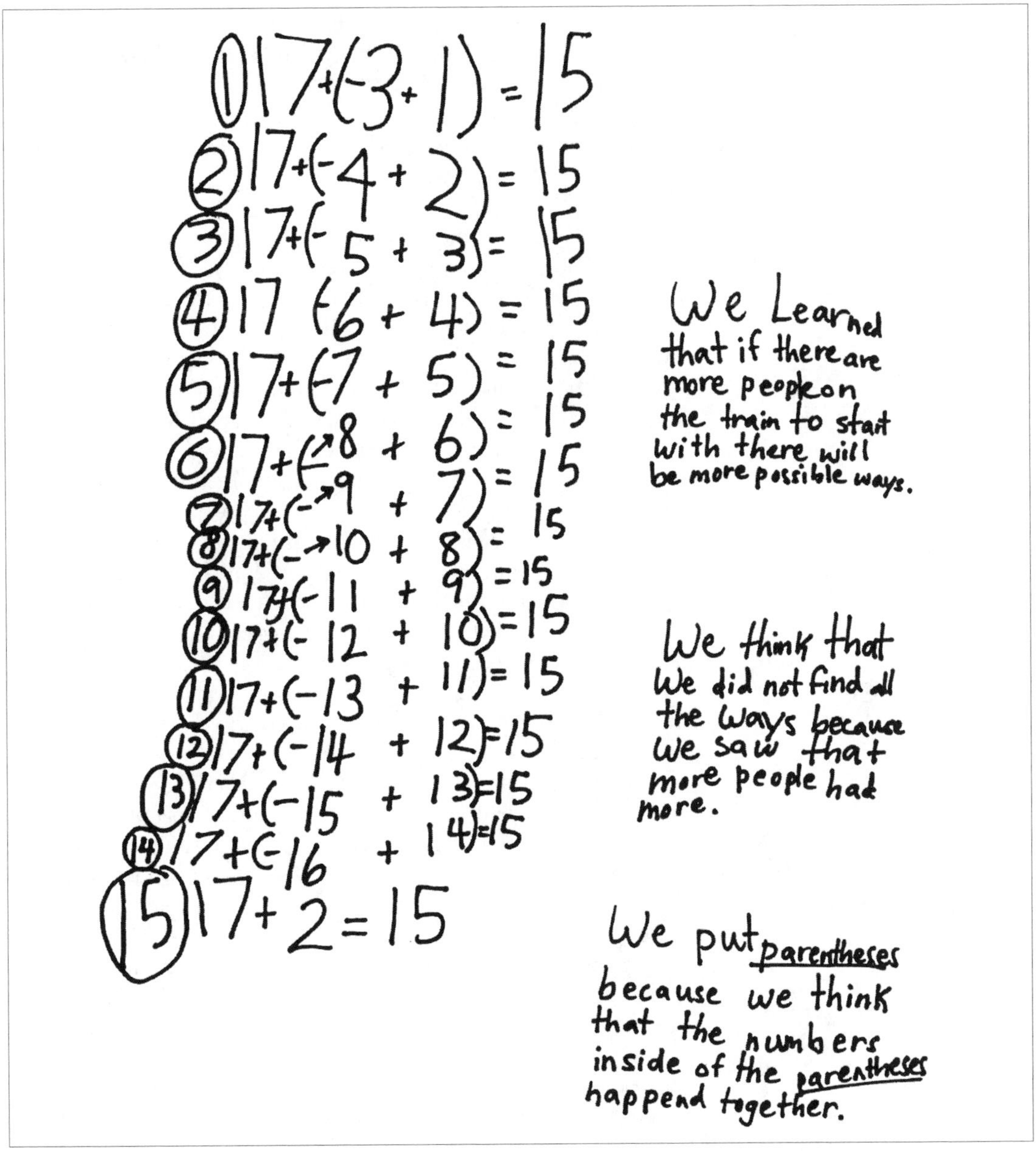

1) $17 + (-3 + 1) = 15$
2) $17 + (-4 + 2) = 15$
3) $17 + (-5 + 3) = 15$
4) $17 (-6 + 4) = 15$
5) $17 + (-7 + 5) = 15$
6) $17 + (-8 + 6) = 15$
7) $17 + (-9 + 7) = 15$
8) $17 + (-10 + 8) = 15$
9) $17 + (-11 + 9) = 15$
10) $17 + (-12 + 10) = 15$
11) $17 + (-13 + 11) = 15$
12) $17 + (-14 + 12) = 15$
13) $17 + (-15 + 13) = 15$
14) $17 + (-16 + 14) = 15$
15) $17 + 2 = 15$

We Learned that if there are more people on the train to start with there will be more possible ways.

We think that we did not find all the ways because we saw that more people had more.

We put parentheses because we think that the numbers inside of the parentheses happend together.

FIGURE 7.4 *Ava and Danielle's Work*

school. But is this model useful for the work middle school students are required to do? And perhaps more to the point, is it useful for algebra? To find out, let's listen in on a Math in the City workshop where elementary teachers are using this model.

TEACHING AND LEARNING IN A PROFESSIONAL DEVELOPMENT SEMINAR

Bill introduces the investigation by telling how he enjoyed hearing a mariachi group on the Lexington Avenue subway. He then poses a set of problems for the group to investigate:

> Mark, Shelia, and Annie want to put on sixty-second comedy shows on Saturday mornings for people who are riding the subway. They decide to start with the F train because it takes them from their school in Brooklyn to midtown, but they aren't sure between which stations to perform. They want cars with enough people but not too many people, because they need room to move around. They decide to collect some information by riding the subway and counting people on some of the cars. Since it's hard to count everyone on a car, they decide to count the number of people who get on and off at each door at each stop.
>
> *Part 1: Testing the theory.* Twenty-six people are in Mark's subway car when it pulls out of 14th street. At 23rd street, people get on and off and after the doors close there are eighteen people in the car. Mark thinks that eleven people got on and nineteen people got off, although he is not sure. He writes $11 - 19$ to record his thinking. What else could have happened; how many people might have entered or left the car? How could you represent the possibilities?
>
> *Part 2: Rockefeller Center.* A few stops later when the car approaches Rockefeller Center there are fourteen people in the car. Mark sees six people get on and three get off, so there should be seventeen people in the car. But there are thirty-two people in the car! Mark realizes he only watched the door next to where he is sitting, but the subway car has three doors! This is going to be tricky. How can Mark list the possibilities? What could have happened at Mark's door (in/out) and what could have happened at the other two doors (in/out)?

The teachers set to work on the first problem. The scenario seems simple enough. Several people write $26 + 11 - 19 = 18$ to record what seems to have happened. Mark could be right. But what else might have happened? One participant uses variables and writes $26 + x - y = 18$.

Two groups decide to make t-charts to record the in/out possibilities (see Figure 7.5). Someone says, "You have to make sure that when you change the ins you do the same for the outs." Bill says, "Okay, talk about how you change them and put that on your posters."

A third group recalls that number lines have been important in the workshop and they create the diagram in Figure 7.6, which they call "linked number lines."

The fourth group remembers the work they've seen Patricia's students do. They make the chart shown in Figure 7.7.

Having represented the combinations that could have occurred in part 1 of the investigation, participants turn their attention to part 2. It isn't very long before they start complaining: "This is going to be a real mess." "I don't see how we could possibly record all of this." Before the frustration leads to disengagement, Bill tells each group, "The whole point here is for the context to push you to find a way to organize what is going on in a reasonable way in a reasonable amount of time."

"You mean do the impossible?" Celia laughs.

Bill smiles but urges the group on. "No, in fact remember our conversation about algebra being about the act of structuring? Here is your chance. Think about how you could represent the possibilities." Bill lets the groups work, and then each one presents their poster.

The first poster includes the equations in Figure 7.7 and an in/out number line similar to Figure 7.6. Discussing part 1 they explain, "If Mark

IN	OUT		OUT	IN
11	19		19	11
10	18		18	12
9	17		17	13
8	16		20	10

FIGURE 7.5   *Two T-Charts for the In/Out Possibilities*

19	18	17	16	15	14
11	10	9	8	7	6

FIGURE 7.6   *Linked Number Lines*

$$11-19 = 10-18 = 9-17, \cdots, = 1-9 = 2-10 = \cdots$$

FIGURE 7.7   *Chain of Equivalences*

wanted to use x and y, he could use x being on and y being off and then all of these equations would work. They are the same as in the chart." The group decides to add two more statements to their poster: $26 + (x - y) = 18$; and $(x - y) = -8$. They explain the second equation as "y is eight more than x."

All the groups appear to understand part 1 of the problem as relating two variables for on and off. They also understand that when you only worry about one door, you can make a complete list of the possibilities. Therefore, Bill turns the conversation to the problem in part 2. He asks Roger and Marcy to share their work, which is shown in Figure 7.8.

Roger and Marcy explain that there are many things that could have happened and that their poster only shows a few of them. They point out the column in the chart indicating that the net change in the end is always +18.

"Look at the top three rows above the line," Roger says. "They are what could have happened if we knew the net change that occurred at each door. And look how complicated it is already! We knew we couldn't list all the possibilities, so we just put some down."

Marcy adds, "Yes, and if you go below the line it gets worse. There are other things that can happen at the different doors; some may go down and others have no change like you see in door three."

Bill asks, "How does your diagram help you know the possibilities?"

Marcy continues, "Well, it really shows it's a mess, lots of things can happen."

"But could you organize it in a way that shows the possibilities without a long list?" Bill probes.

"You could just write $3 + 7 + 8$ for the top three lines," Roger replies. "It's just the totals at each door."

"Yeah, but that's just net change, not all the changes," Marcy cautions.

"But what else can we do? This whole thing would be too big otherwise," Roger responds.

Eventually the participants agree that it would be helpful to keep track of the possible net change at each door, knowing it must be related to the net change at the other doors (and always keeping in mind that there are numerous possible scenarios at each door).

FIGURE 7.8
Door–by–Door Possibilities

$$3 + (x + y) + 14 = 32$$

Door 1	Door 2	Door 3		
$6 - 3 = 3$	$10 - 3 = 7$	$12 - 4 = 8$	$+14 =$	32
$7 - 4 = 3$	$11 - 4 = 7$	$13 - 5 = 8$	$+14 =$	32
$8 - 5 = 3$	$12 - 5 = 7$	$14 - 6 = 8$	$+14 =$	32
$5 - 2 = 3$	$17 - 1 = 16$	$0 - 1 = -1$	$+14 =$	32
$4 - 1 = 3$	$15 - 0 = 15$	$15 - 15 = 0$	$+14 =$	32

Another group is then invited to discuss their poster, which is shown in Figure 7.9. They have drawn a rectangle to represent the subway car with three doors. They have assumed Mark is correct in his count at the door he was watching and have written +6 − 3.

Cynthia speaks for the group. "We decided to make our work easier by assuming that Mark had it right at the first door. Six people got on and three got off. This way we only had to think about the other two doors, and we knew that whatever happened there we had to have a net gain of 15. There are lots of ways to do this. But the point is to keep track of the net change at each door, because that is the best way to sort the possibilities." She displays a chart showing some of the possibilities (see Figure 7.10).

Continuing, she points to the rows of the poster where she has 0 + 15, 1 + 14, and so on. "If nobody gets off, then this is all that can happen. But even if nobody gets on or off at door 2, there are lots of possibilities for door 3 and we have this in our second column." She writes 0 + (16 − 1), 0 + (17 − 2), 0 + (18 − 3).

"So explain your strategy to represent all possibilities," Bill says. *He's pushing for the big idea.*

"Well, as I said, we have a t-chart listing all the net change possibilities. These are all of them, since no more than fourteen people will get off and Mark saw three get off. The net changes at the other two doors have many possibilities, and these have to be listed next. But we didn't make a full chart of these, only some examples. The point is that you find the relation between the net changes first and then think about the possible configurations for each

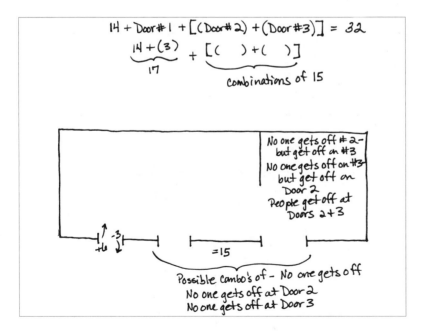

FIGURE 7.9   *Net Change at Each Door*

net change. The only limit is that no more than fourteen people are going to leave at Rockefeller Center."

WHAT IS REVEALED

Cynthia and her teammates have constructed a big idea—*net changes can be used as quantities*, not only as the result of two operations. Moreover, when thinking this way, each net change includes all possible equivalent pairs of operations—for example, $-3 = 0 - 3 = 1 - 4 = 2 - 5 = 3 - 6 = \ldots$ and so on. Each net change represents a list of possibilities, which in this context is limited by the number of people who can leave the car (and, of course, the car's space limits). The context has provided the opportunity for net changes to become objects to work with, the operations governed by properties of the counting numbers. By doing this, Cynthia's group is now working with integers. The calculations they do with integers will then be accomplished using one of these representations for net change (for example, $11 - 19$, for a net change of -8).

In this context, integers are mental objects consisting of equivalent differences leading to the same net change ($11 - 19 = 12 - 20 = \ldots$), and operations with integers can then be carried out by using these mental objects. For example, if we want to add $4 + -8$, the model allows us to consider these numbers as net changes and combine possible equivalent expressions, such as $(5 - 1) + (12 - 20)$, which is $17 - 21$, a net change of -4.

The subway context allows Bill to raise the issue of net change and the equivalences that underlie net changes. In a car with three doors, assuming Mark's count is correct at one door, the possibilities at the two other doors are classified by first describing the relationship between the net changes (what Cynthia's group represented as $x + y = 15$); then each net change represents a collection of equivalent pairs of operations. But can all operations with integers be modeled as net changes? What if we need to subtract?

FIGURE 7.10
*Possibilities of
Net Change*

x	y
-11	26
$\cdots$	$\cdots$
-1	16
0	15
1	14
$\cdots$	$\cdots$
7	8
$\cdots$	$\cdots$
26	-11

Bill again uses the subway context. "A subway car is going from station A to station B to station C." He draws a figure with arrows between the letters A, B, and C. "Suppose that 24 people were on the car when it left station A and 36 people arrived at station C. What happened at station B?"

$$A \longrightarrow B \longrightarrow C$$

"There is a net change, +12," Hugo replies confidently.
"How did you know?"
"Because 36 − 24 = 12."
Bill records:

24	?	36

$$A \longrightarrow B \longrightarrow C \quad 36 - 24 = 12$$

Then he asks, "What if the next time we have this?" He writes:

?	24 (net gain)	36

$$A \longrightarrow B \longrightarrow C$$

"It's the same equation as what Hugo said," Marcy replies quickly, adding, "It is still 36 − 24 = 12. Twelve people leaving the station at A."
"Why?" Bill persists.
"Well, if the net gain was 24 at station B, to find how many left station A, we have to remove 24 from 36."
"Okay, but how come Hugo used the same subtraction expression for the first context?"
Marcy hesitates a moment, then smiles. "Well, that one is difference, because you are growing from 24 to 36—oh, yeah, one is the difference model and the other is removal. That's neat!"
After making sure participants understand why Hugo's answer was based on difference and Marcy's on removal, Bill presents the next problem in the string. "How about this one?" He writes:

?	−24 (net loss)	36

$$A \longrightarrow B \longrightarrow C$$

"It's 60!" the group blurts out.
"But what's the equation?" Bill responds. Tom suggests hesitantly, "I think we have to use Marcy's approach. It should be 36 − −24 equals the amount. That's 60 isn't it?"
Bill records 36 − −24 = 60 and asks the group to discuss why they think this equation is correct in this context. Alice suggests that the removal is like going back in time: You are finding out what would have happened if a net change of −24 hadn't occurred.

With this string of related problems, the group has taken another step. Net changes have been given a new status: They can be operated on with counting numbers. When Tom suggests that the context gives $36 - -24 = 60$, not only has he calculated the value but *he has expanded the number system to include negative numbers.* True, he's hesitant; he's using Marcy's approach, which seems strange at first, but because he stays in the context and thinks about the situation, the equation makes sense, so he accepts it. This is a feature of the construction of integers—the operations work compatibly with the operations with counting numbers we are familiar with—and a good model will support this realization. Because they must work compatibly with the operations of the natural numbers, the addition and subtraction "rules" must be what they are. But this realization comes much later; for now, what is most important is that the operations make sense.

There is another way to represent subtraction as removal, and it is linked to the equivalences that are used when thinking about net change. Suppose six people board the subway and fourteen get off, but six of the fourteen jump back on at the last second before the doors close because they realize they are at the wrong station. In this case we are now losing –6 of the original net change. If we represent the first change as 6 – 14, we may then remove the –6 by removing (0 – 6). The resulting equation can be written as $(6 - 14) - (0 - 6) = (6 - 8)$, because $14 - 6 = 8$. The six that jump back on really are removed from the fourteen people getting off. Thinking about net change as a quantity like this can be represented as $-8 - -6 = -2$. This approach also works in other contexts and models used to develop integers—equivalences of + and – chips, for example (see Figure 7.11).

Adults in math methods courses like the representation in Figure 7.11 because "at last" the rule they memorized seems to have an explanation. But curriculums using chips can have serious drawbacks for children, because teachers often use the model to teach subtraction with integers procedurally and learners do not construct the role of equivalence. In short, a manipulative cannot do the teaching—the construction of mental objects requires contexts and problems rich enough to raise the important issues. The contexts involving subways have prompted learners to grapple with equivalence and net change simultaneously.

FIGURE 7.11
Subtraction as Removal with +/– Chips

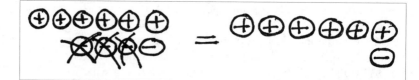

The students in Patricia's class took net change into account in order to understand how the number of people on a subway car changes after a stop. They found that two numbers need to be considered: the number that get on and the number that leave. Moreover, they found that these numbers are related if the net change is known.

Operating with integers is not yet on the horizon for these second graders but equivalence is and it lays the foundation for the work in later grades, where, as with the teachers in Bill's workshop, equivalences in net change, such as $0 - 3 = 1 - 4 = 2 - 5 = \ldots$ must give rise to a new mental object—the negative number -3, for example.

The teachers in Bill's workshop created Freudenthal's *mental objects* and then realized that these objects are handled by *mental operations*. The subway context—considering all possible pairs whose difference gives the same net change—led to the big idea that net changes can be considered quantities and that these new objects can be operated on and with. As Pascal said, "We are usually convinced more easily by reasons we have found ourselves than by those which have occurred to others."

8 | COMPARING QUANTITIES AND RELATIONS

Mathematicians do not study objects, but relations between objects.

—Henri Poincare

Each problem that I solved became a rule, which served afterwards to solve other problems.

—René Descartes

First-year algebra students learn to solve systems of equations like these:

$$3x + y = 40$$

$$4x + 2y = 58$$

A number of techniques are traditionally taught for finding the solution, such as eliminating a variable or isolating a variable and then substituting one equation into the other. If left to their own devices, many students try to solve the system by guessing and checking, find it extremely tedious, and give up.

Try thinking about the following context instead (answer the two questions before you read further):

- Suppose 3 suckers and 1 gum ball cost 40 cents, while 4 suckers and 2 gum balls cost 58 cents.
- What happens to the price when you add one sucker and one gum ball to your purchase?
- What is the price if you decided to only buy 2 suckers and 0 gum balls?

Adults often have more trouble with these questions than children do, because they try to use symbolic procedures without understanding the underlying algebraic ideas.

Most children do something like the following. Instead of thinking about individual values, they notice that it was 18 cents more to buy 4 suckers and 2 gumballs than it was to buy 3 suckers and 1 gumball, and thus they reason that the cost of one sucker and one gumball must be 18 cents. To answer the second question, they remove the cost of a sucker and

129

a gum ball from the cost of 3 suckers and 1 gum ball, determining that 2 suckers and 0 gum balls must be 40 − 18 = 22 cents. Note that from here the individual price of a sucker or a gum ball can be readily determined—11 cents for a sucker and 7 cents for a gum ball.

The goal of these questions is different from the goal of asking for a solution to the values of x and y—the context and questions deliberately raise the issue of how the quantities are related. They link the prealgebraic strategies involving number developed earlier in this book to problems involving several unknowns. Contexts that have the potential to suggest comparing quantities provide students opportunities to develop prealgebraic abilities, such as reasoning with unknown quantities, using and generalizing relations, and developing notation to support such reasoning. They are important for the development of algebra because they focus attention on relationships between combinations and equivalence. Rather than teaching elimination methods as is done in procedural algebra courses, the goal is for students to reason with the quantities in context and develop the big ideas that will be needed later for making sense of elimination methods. This is Poincare's point when he says, "Mathematicians do not study objects, but relations between objects." The equations $3x + y = 40$ and $4x + 2y = 58$ have independent meaning, but when they come together in this context the relations between them are crucial.

Research has shown that students can use their knowledge of a context to think about relationships between quantities and that the strategies they develop are useful in solving systems of equations with two, three, or more variables (Meyer 2001). Working with the stories and pictures in the *Mathematics in Context* series of curricular units, students imagine exchanges and equivalences to determine values and relationships in problems like those in Figure 8.1.

Motivated by our success with using *Mathematics in Context* materials, we studied contexts that would develop these same abilities but would further lead students to represent their comparisons and equivalences using number lines and double number lines. To set the stage for a discussion of this work, let's listen in as Bill presents a few problems to a class of sixth graders.

TEACHING AND LEARNING IN THE CLASSROOM: MIDPOINT PROBLEMS

"I'm thinking of two numbers," Bill begins. "When I add them I get 20 and when I subtract them I get 10. Can you find the numbers?" The students consult with a nearby partner for a few minutes.

"I think they both have to be ten, because ten plus ten is twenty," Marcy offers.

"Okay, but what happens when you subtract? Did you get ten?" Bill replies.

Comparing Quantities
_____

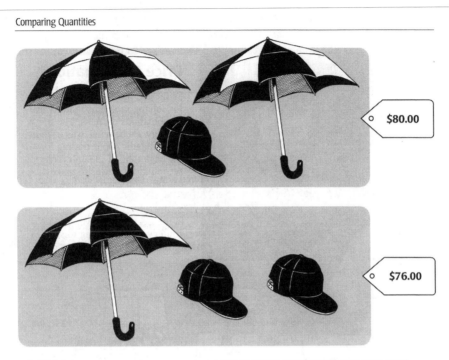

4. Which is more expensive, a cap or an umbrella? How much more expensive is it?

5. Use the two pictures above to make a new combination of umbrellas and caps. Write down the cost of the combination.

6. Make a group of only caps or only umbrellas. Then find its price.

7. What is the price of one umbrella? one cap?

Britannica Mathematics System

FIGURE 8.1   *Comparing Quantities: Hats and Umbrellas*

Marcy shakes her head no and thinks some more. Her partner, Tamar, offers a different idea. "I think one number has to be fifteen and the other five, because fifteen plus five is twenty and fifteen minus five is ten." Most students have used guess-and-check to arrive at the same answer, and Bill represents the solution on an open number line (see Figure 8.2). "Here is fifteen, and if we jump up five we get to twenty and if we jump back five we get to ten." Then he challenges, "Okay, I have some new numbers. When I add them I get 30 and when I subtract them I get 10."

"Twenty and ten," Hugh blurts out quickly. As most of his peers nod in agreement Bill says, "Okay. That was too easy. What if the numbers add to thirty, but their difference is fourteen?" This time the class is silent. Bill asks the group to take a minute and think about it.

Some students continue to guess and check, counting on their fingers, "Let's try seven plus twenty-three, then it's thirty, and counting back we get . . . sixteen. No, let's try six plus twenty-four, going back . . . wrong way. Okay, let's do eight plus twenty-two, and we get . . . fourteen. Fourteen—that's it! The numbers are eight and twenty-two."

Bill consults with Hugh. "You were so quick to see that the numbers were ten and twenty just a moment ago. Could your strategy there help you now?"

"Well, I went to the middle," Hugh says. "Twenty was the middle of ten and thirty."

"Could you do that now?"

Hugh returns to work, beginning by drawing an open number line.

After a few minutes, Bill asks for everyone's attention and solicits some strategies. Most students have found that 22 and 8 are the solution, and their strategies follow basic patterns. Some guess, check, and revise, getting to the solution in one or two tries. In the literature these approaches are referred to as *arithmetical* strategies (van Ameron 2002). Other students realize that because 10 and 20 don't work anymore they can adjust by adding half of the difference between 10 and 14; adding 2 to 20 they get 22. Or they take a different initial estimate and adjust by taking into account how far off they are. These are *prealgebraic* strategies—the adjustments are based on the numerical relationships between the numbers they tried and the ones they want to get (van Ameron 2002). Bill notes the evolution of strategies discussed thus far and asks Hugh to report on his drawing (see Figure 8.3).

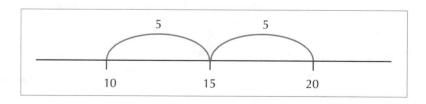

FIGURE 8.2
Midpoint Problem

"Well, here is fourteen and here is thirty on the number line. One of the numbers we want to find is in the middle because you have to go up and down the same amount. So I did thirty minus fourteen equals sixteen, and I had to go up eight from fourteen. That gives me twenty-two, and the other number has to be eight."

Hugh has used an algebraic strategy. It is a generalizable process that can be used to find x and y if both x + y and x − y are known. Represented symbolically, Hugh has subtracted (x + y) − (x − y) and found 2y. From 2y he gets y by dividing by 2 and then he finds x by adding y to x − y, the smaller number. His diagram is essentially the same as Bill's. Although he is not using the symbols we've used to represent his thinking, Hugh is reasoning with relationships rather than computation—this is early algebra.

Other students also use algebraic strategies. Connie and Maria add 14 and 30 to get 44 and take half of that, also arriving at 22. This strategy also works. Bill asks students to discuss in pairs why the larger mystery number is precisely at the midpoint and why this is the same as adding the two numbers and dividing by two. Connie shares with the class, "You see, the bigger number has to be in the middle, so it has to be the average of the two numbers." *Many students go through school learning to calculate averages without understanding this interpretation on a number line.*

Once this representation of averaging is understood, Bill continues, "We have seen Hugh's strategy and Connie and Maria's strategy. Here is what I would like you to do. I have two problems for you to work on in your groups. They are presented this time in symbolic form." Bill writes the following problems:

Problem 1: $x + y = 151$
$\qquad\quad x - y = 49$
Problem 2: $z + w + w = 151$
$\qquad\quad z - w = 49$

The first problem is a midpoint problem like the earlier two. This time most students move beyond guess-and-check and represent their work on an open number line. Since the difference between 49 and 151 is 102, and half of 102 is 51, that gives y. To find x, many students jump up 51 from

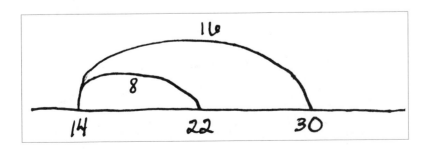

FIGURE 8.3
Hugh's Number Line

49 and realize that $x = 100$. A few students use Connie and Maria's averaging strategy.

The second problem is trickier, but some students generalize the ideas from the first problem and break the distance between 49 and 151 into three equal pieces of length $w = 34$ to see that $z = 49 + 34$. The number line is becoming a tool to compare quantities and for representing algebraic relationships.

WHAT IS REVEALED

Problems in which the sum and difference of two mystery numbers are known are called *midpoint problems* because the larger number occurs halfway between the two known numbers. In this particular lesson, Bill first uses an open number line as a representational tool with an easy version of the question (the numbers were both landmarks), because he wants students to try to use the number line as a tool for subsequent investigations. That 15 is halfway between 10 and 20 is pretty easy for Hugh when he uses 20 as the midpoint between 10 and 30. When the students find the answers readily, Bill pushes them to think about the model and the midpoint in solving the mystery when the sum is 30 but the difference is 14—a harder version. These problems, represented on the number line, prompt students to use the model with the more symbolic versions. Although many students can solve a midpoint problem by generalizing their strategies abstractly (working with numbers in their heads), the number line representation is an essential tool in the more general case illustrated in Figure 8.4. It will also become an important representational tool in the investigation that follows.

BACK TO THE CLASSROOM: BENCHES AND FENCES

Because Bill wants his students to continue to compare quantities and make further use of number lines and double number lines as tools in their work, he brings back the frog context (see Chapters 4 and 6).

FIGURE 8.4
Generalized Midpoint
Problem

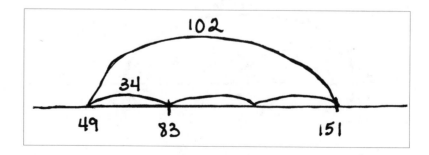

Let me tell you a bit more about the frogs. One day they decide to have a jumping contest. But there are lots of frogs who want to watch, so they decide to set up benches on each side of their jumping tracks. They have a twenty-eight-foot track and a forty-two-foot track. They have decided to bring in benches from their storeroom and place them along both sides of each track end to end so the benches line the track lengths exactly. The benches in the storeroom are of two different lengths. One size is six feet long and the other size is eight feet long. How can the frogs line the two tracks with benches that fit exactly?

After a short conversation ensuring that the context is understood—the requirements that the bench lengths add up to the length of the track and that no bench can be cut—Bill poses two specific questions for them to investigate:

- How many six-foot benches and how many eight-foot benches are needed in order to exactly line both sides of each track?
- Are there other possible choices of six- and eight-foot benches that can be used?

"We can't make a length twenty-eight using just eight-foot or just six-foot benches because six and eight don't go into twenty-eight," Thomas declares in exasperation.

Alyssa offers a possible solution. "But we can mix them up. What if we use three sixes?"

"Okay, that's six, twelve, eighteen." Thomas marks each jump of 6 on his drawing of the twenty-eight-foot track. "But that leaves ten more. That can't work either. Can we cut a bench?"

Overhearing, Bill reminds them that the benches cannot be cut. "No, you can't cut the benches. Alyssa said you could mix them up. Are there other ways to do that?"

"If we do two sixes, that's twelve, which leaves . . . sixteen. That's two eights. Hey, two sixes and two eights work!" Thomas and Alyssa have found a possible solution for the first track.

John, Meg, and Marcy are working on the forty-two-foot track. "If we have four eights, that's thirty-two, and if we have two sixes, it's twelve, so together that is, um, forty-four. It's too much."

"But it's only two too much, so let's switch an eight and a six." *Meg offers a prealgebraic strategy in contrast to the previous arithmetical guess-and-check strategy.*

"Yeah, that's a good idea. That cuts it back to forty-two, and we have three eights and three sixes," John declares with triumph.

Marcy appears perplexed. "But I got seven sixes for forty-two."

"But that's not what we got. Can we have two answers?"

"Sure you can," Bill reassures them. "But when you do, it is important to think about how they are related. Maybe Meg's switching strategy can help you think about this."

The numbers in this scenario have been chosen deliberately. There is a unique way to build each of the twenty-eight-foot lengths (2 six-foot and 2 eight-foot benches can be put together for each side of the track). However, there are two possible ways to build each of the forty-two-foot lengths, either with 7 six-foot benches or with 3 six-foot benches and 3 eight-foot benches.

Although most students easily find an answer, Bill pushes them to think about finding other possibilities and to verify that they have found them all. The big idea of equivalence emerges when students are able to see that 4 six-foot benches cover the same distance as 3 eight-foot benches. This leads to the important algebraic strategy of substitution; 3 six-foot benches and 3 eight-foot benches can replace 7 six-foot benches on one side.

At first Bill just listens as the students work, wanting to understand their ideas and strategies (students' initial ideas are always the beginning of a good conference). Many students working on this problem draw each attempt on a representation of the track. Although guess-and-check is common initially, trial-and-adjustment is an important strategic advance. This type of thinking, illustrated on the representation as the decomposition $8 = 6 + 2$, is a precursor to the formal algebraic operations students will use later. Bill's final comment leaves the possibility that students may see that four sixes is equivalent to three eights with an eye toward using equivalence as a tool for examining the possibilities.

To further facilitate construction of equivalence, the context specifies placing benches on both sides of the track. Many students use a rectangle and mark off lengths on opposite sides of this rectangle to represent the sections they are using on the two sides of the track. Others use arithmetic (see Figures 8.5a and b). Students who have constructed the equivalence of four six-foot benches with three eight-foot benches are able to show this equivalence on their representations.

BACK TO THE CLASSROOM

To encourage construction of this equivalence Bill adds one last condition to the problem. "While you are working, let me tell you one more part of the story. When the frogs went to the storeroom, they found that they had only seventeen six-foot benches and nine eight-foot benches. Will these amounts work? What should they do? How many six-foot benches and how many eight-foot benches are needed to line both tracks? Are there other possible choices of six-foot and eight-foot benches that could be used? How do you know you have them all?"

After students have had time to consider this extension of the problem, Bill has them prepare posters (see Figures 8.6a and b) and convenes a math

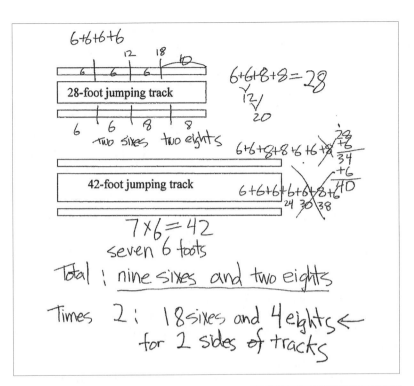

FIGURE 8.5a
Work on the Bench Problem

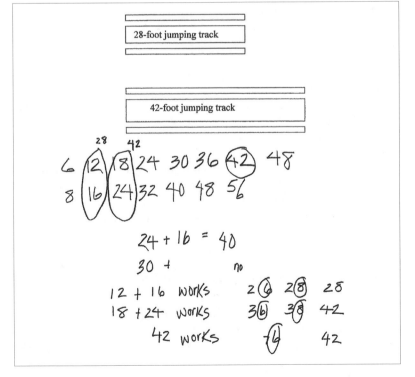

FIGURE 8.5b
Work on the Bench Problem (continued)

congress to discuss a few of them. He focuses on how they know that the three possibilities are all the possibilities that can be found.

"Let's start with you, José and Lupé. You noticed something about replacing eights by sixes. Tell us about it."

"For the small track, the twenty-eight, we knew that two sixes and two eights worked," Lupé explains. "If you try to replace any eight by a six it gets two smaller, and if you do it twice it gets four smaller and then you are stuck. So there aren't any more ways to do it."

José nods and continues, "But with the bigger track we knew that three sixes and three eights worked, and if you take three eights out and put in three sixes, then you are six too small. You can fix that by putting in another six. So seven sixes work."

Lupé adds, "And six times seven is forty-two so that also shows why it works."

Bill pushes for proof. "How does that show that we have all the solutions?"

Lupé responds, "Because we can't go any smaller or bigger."

Bill asks for a paraphrase to see what the other students are thinking. "Can any of you put what Lupé is saying into your own words? Tara?"

"Well, three sixes and three eights work for the long track. If we want to do it a different way, say more sixes, we put a six in and take an eight out, but then it is two shorter. So we do that three times and then have six, I mean seven, sixes. But you can't take any more eights out so that's all of them. That's why you did them all."

Alfonso is puzzled. "But what about if you want fewer sixes?"

Tara explains, "But if you take sixes out and put eights in, it gets bigger."

Alfonso isn't satisfied. "Yes, but if you take a six out and don't put an eight in, it gets smaller."

Tara tries to convince him. "Yeah, but then it gets too small because there are only three sixes and you can't put eights back in to fix it. I think you have to put in four sixes and take out three eights to make it work."

Tara has just expressed an important algebraic idea: Equivalent expressions can be substituted. Bill says, "Did you hear what Tara just said? Talk about Tara's last statement with your partners for a minute. Does her idea help us to know if we have found all combinations of sixes and eights that make forty-two?"

WHAT IS REVEALED

The purposefully chosen limitation on the number of benches requires that students use both solutions for the forty-two-foot track and thereby consider equivalence. Bill begins by asking two students to share a general

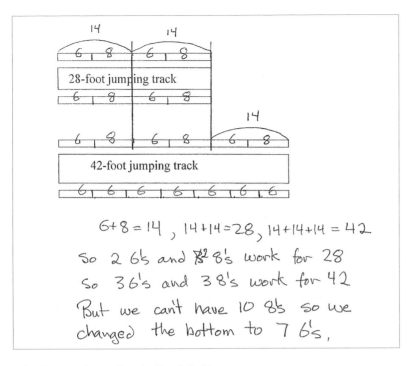

$6+8=14$, $14+14=28$, $14+14+14=42$

So 2 6's and 2 8's work for 28

So 3 6's and 3 8's work for 42

But we can't have 10 8's so we changed the bottom to 7 6's,

FIGURE 8.6a   *Posters from the Bench Problem*

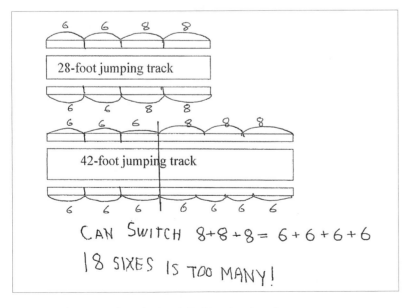

CAN SWITCH $8+8+8 = 6+6+6+6$

18 SIXES IS TOO MANY!

FIGURE 8.6b   *Posters from the Bench Problem (continued)*

observation they have made. This implicitly suggests that part of doing mathematics is communicating and justifying thinking to a community of other mathematicians. Discussion is welcomed and flows naturally.

Bill directs the conversation toward the central issue of justifying the claim that all the possibilities have been found. Asking for clarification and paraphrasing ensures that students understand one another's ideas and can discuss them. Students defend their thinking—Bill doesn't do it for them.

The equivalence of four sixes and three eights has come up as a strategy for examining the possibilities. This is explored in greater detail as the congress proceeds. By the end of the congress the students have made convincing arguments that you cannot have more than two possibilities for the track of length forty-two feet.

The context of placing benches on both sides of a jumping track leads students to represent combinations in parallel—an emergent double number line. This facilitates representing the equivalence of three eights and four sixes on a double number line as in Figure 8.7 and reveals further relationships. The second six ends in the middle of the second eight. Also the first eight exceeds the first six by two, so it is one-third of the way to the end of the next six. These relationships also illustrate Lupe's idea that when you replace an eight by a six it gets two smaller—an exchange that requires addition to form an equivalence.

Lupé and José's exchanges also prove that all combinations have been found. If you start with a combination that works, you cannot remove one or two eights and replace them with sixes because you will be two or four short. The equivalence of three eights with four sixes needs to be used, proving that all combinations have been found, which is more efficient than checking all possibilities. This idea can also be represented nicely with another powerful model—a combination chart, which is where Bill goes next.

BACK TO THE CLASSROOM: THE COMBINATION CHART

Bill continues with the frog tale. "The frogs now want to build a fence to enclose a rectangular jumping arena. The arena is fifty-two feet by sixty-six feet. They can buy fencing in six-foot and eight-foot lengths. What are all the possible choices of six-foot and eight-foot sections of fencing to go all around the arena? Like before, they cannot cut these fence sections and they cannot bend them around a corner. One six-foot section will come with a gate. Help the frogs prepare a shopping list for the different possibilities so they can buy enough sections of fence."

Because Bill is using larger numbers this time, the students are pushed to go beyond guess-and-check and use the equivalence and substitution strategies they developed previously. But because there are more possibilities this time, students are also challenged to organize their work and record possible exchanges. Again, the numbers in this scenario have been chosen purposefully. Each length is twenty-four feet more than in the previous scenario, so students could obtain initial solutions by adding four sixes or three eights to the solutions for the benches.

At first, students don't notice this and start over, finding an initial combination and modifying it by using equivalent exchanges to find the other possibilities. Eventually a variety of strategies emerge as students work. Some of the posters they create for the congress are shown in Figure 8.8. Marcy and José create a sequence of diagrams with possibilities determined by arithmetic (Figure 8.8a), while Clarissa and Juan create number lines for each fifty-two-foot and sixty-six-foot length, with solutions shown of the substitutions of four sixes for three eights on double number lines (Figure 8.8b). Other students work with charts of possible values, but Maria and Roberto list all the multiples of 6 and all the multiples of 8 and then add various combinations of these multiples together to see which ones add to 52 (Figure 8.8c). Sam and June create a chart showing one possibility (such as 6 six-foot pieces and 25 eight-foot pieces) and then note the corresponding numbers that arise by making successive exchanges (Figure 8.8d and e).

The next day, Bill organizes a gallery walk to examine the posters and then convenes a math congress. Rather than discussing the posters (by this time, based on their experience with the bench problems, the students are convinced they've found all the possibilities), he introduces the combination chart. He begins by explaining that often mathematicians solve problems by stepping back and considering a more general question than originally posed. This allows them to think more deeply about the relationships involved instead of the specific procedures to get the answer. He tells the class that this is what they will be doing today. Instead of going through the various approaches to build fences fifty-two and sixty-six feet long, the class will construct a combination chart, which will, among other things, include different ways to combine six-foot and eight-foot lengths to make a total of fifty-two or sixty-six feet. However, the organization of the chart will be of special interest. Every "move" from one entry to another corresponds to adding or subtracting lengths. In this way, all of the exchanges considered on previous days can be represented on the chart.

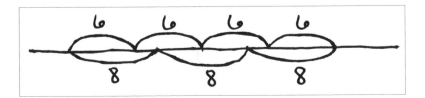

FIGURE 8.7
The Equivalence of Four Sixes and Three Eights on a Double Number Line

FIGURE 8.8a
Student Work on the
Fence Problem

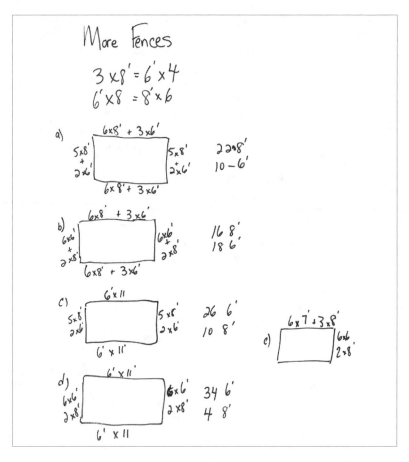

FIGURE 8.8b
Student Work on the
Fence Problem
(continued)

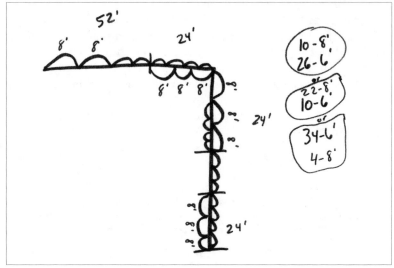

142

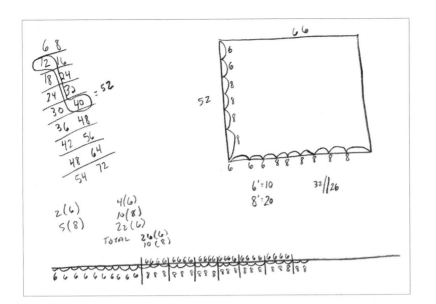

FIGURE 8.8c
*Student Work on the
Fence Problem
(continued)*

6 8
12 16
18 24
24 32   = 52
30 40
36 48
42 56
48 64
54 72

2(6)   4(6)
5(8)   10(8)
        22(6)
TOTAL 26(6)
        10(8)

6' = 10   32/26
8' = 20

66

6
6
8
8
8
8
52
6 6 6 8 8 8 8 8 8

6' 8'
1. 6   8
2. 12
3. 18   16
4. 24   24
5. 30   32
6. 36
7. 42   40
8. 48   48
9. 54   56
10. 60

4 - 6ft = 3 - 8ft.

66
6' | 8'
11 | 0
7 | 3
3 | 6

52
6' | 8'
6 | 2
2 | 5

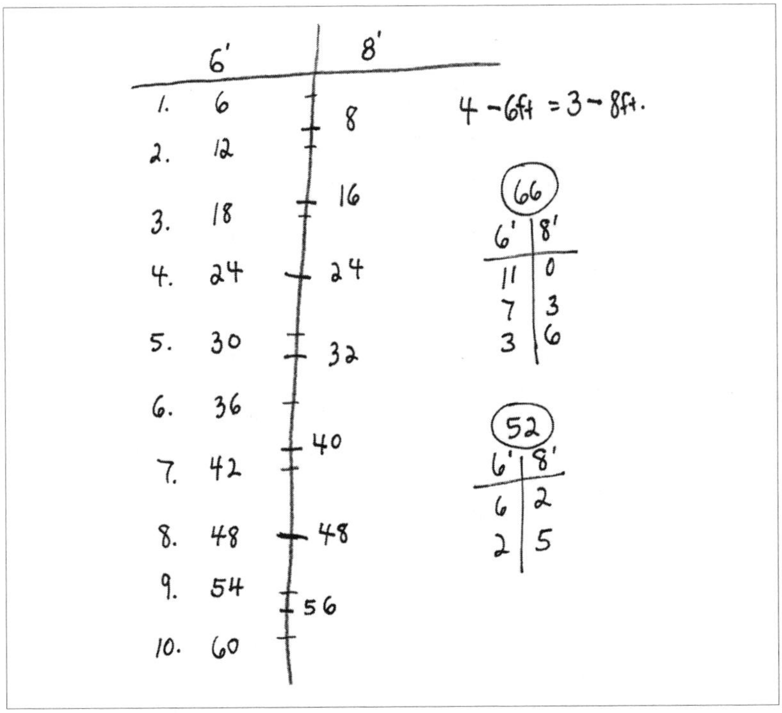

FIGURE 8.8d
*Student Work on the
Fence Problem
(continued)*

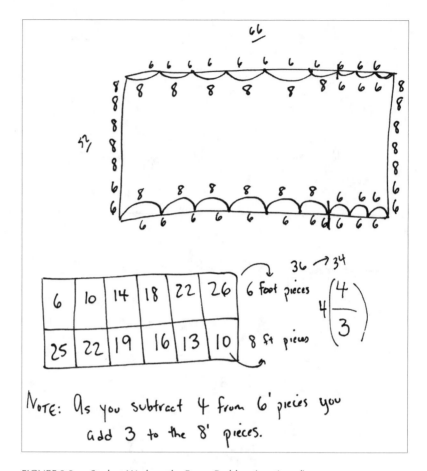

FIGURE 8.8e   *Student Work on the Fence Problem (continued)*

"Clarissa, you were talking about what if there were no eight-foot sections. Then there would be some problems for fence makers," Bill begins. "Tell us about that and we'll start there with a chart."

"Juan and I were saying that if the store ran out of eight-foot sections, it would be bad because you could only do multiples of six, like six, twelve, eighteen, and so on," Clarissa says.

"Well, that might happen, right? And what if you wanted to build a thirty-foot fence?"

"That would be okay because you are lucky." Juan smiles. "You could just use six-foot sections. You'd need five of them."

"So let's start our chart by putting that information down." Bill begins the combination chart shown in Figure 8.9. "Does this look okay, Clarissa and Juan?" They nod their approval. "But what if we had the opposite problem? Say the store only had eight-foot sections?"

6	12	18	24	30	36	42	48
1	2	3	4	5	6	7	8

Number of six-foot sections

FIGURE 8.9 *Beginning the Combination Chart*

"Then we'd have to use only the eights. Like sixteen, twenty-four, and so on."

"Okay," Bill replies. "But I'm going to write them as a column instead of a row. Like this." *Bill models the importance of looking at both extremes, an important strategy for mathematicians* (see Figure 8.10). "Any thoughts on why I'm doing it this way? Why rows and columns? Talk to your partner for a minute and see if you can find out why."

After a few minutes, Bill calls on Rosie. "Rosie, what's your theory?"

"I think you made a big rectangle with rows and columns so you can put in all the other numbers," Rosie declares confidently.

Sam is perplexed. "I don't get it. What other numbers?"

"So we can combine them," Rosie explains. "Look. If you have one six and one eight, that's fourteen and you can put it here," pointing to the cell above the six and to the right of eight.

"Oh, you mean we should just add them up and fill them in, like adding twelve and eight and putting in twenty in the next spot?" Sam asks.

"We could," Bill agrees. "Would a chart like this be helpful?"

"I guess. Customers could just read off the chart then."

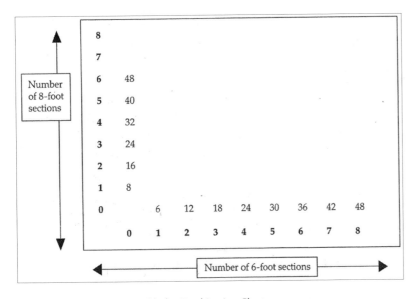

FIGURE 8.10 *Continuing with the Combination Chart*

Bill continues building the combination chart. Then he encourages the students to examine the layout. "Is this organization helpful? Why? What might go in next?"

"We could just go across and add six each time," Ramiro offers. "That might be easier."

"Explain what you mean."

Ramiro continues, "Next to the thirty-four. It would be forty, then forty-six." (See Figure 8.11 for Ramiro's numbers.)

Bill encourages everyone to reflect on this important idea. "Why does Ramiro's strategy work?"

Charlene says tentatively, "I think it is because each time you are adding another six-footer." And then with more confidence, "Yeah, that's it. Each time there is one more six-foot piece so the fence grows by six."

Bill now has students fill in their own copy of the chart. While they work he asks them to consider the following questions:

- What happens when you go up two rows? Why do the numbers on the chart go up by 16?
- What happens when you go up one row and then to the left one column?
- What happens if you go down two rows and over three columns?
- The number 30 is on the chart twice. What does that mean? Why did that happen? What kind of exchange is happening?
- Where do 52 and 66 appear? Why there? Are those solutions on the posters?

The students find these and other relationships and discuss how they help them fill out the chart. They examine the chart to verify that they

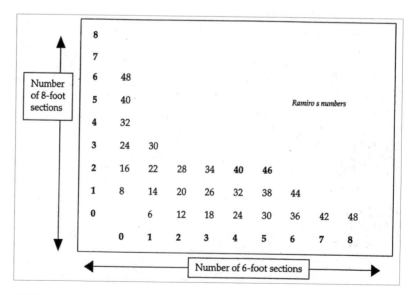

FIGURE 8.11   *Adding Ramiro's Numbers*

WHAT IS REVEALED

Thinking about these questions focuses students' attention on the relationships encoded in the chart. They note that going up one and right one is the same as adding 14. But this makes sense, because they are adding one six-foot and one eight-foot fence piece, so it is the same as adding 14. Similarly going right one and up two is the same as adding 20. While working on the chart, some students have complained that filling it out is laborious, so Bill teases them. "I want you to do this, but I also want you to be lazy. You have to think about what is happening on the chart, how you can easily move up, down, diagonal, or like the chess piece that looks like a horse. Then you can develop a strategy that works in a lazy way for you."

Students who are used to being assigned repetitive problems may indeed view the chart as yet one more: They laboriously add numbers of sixes and eights to fill in each box, without seeing the relationships. However, creating the combination chart builds on ideas developed in the previous investigation with the benches, and it is important to encourage students to take advantage of them. As Descartes noted, "Each problem that I solved became a rule, which served afterwards to solve other problems." This is exactly the idea here—the combination chart encodes many rules that will help in situations yet to arise.

BACK TO THE CLASSROOM: COMPARING QUANTITIES AND SYSTEMS OF EQUATIONS

Bill now wants his students to use the number line as an approach to more challenging problems that require comparing quantities. He tells the following story:

> The frog-jumping Olympics opens with the pairs competition. Each pair gets two jumping sequences. The winner is the pair that has the longest distance when their two jumps are combined.
> **Team #1** (Huck and Tom): When Huck jumps three times and Tom jumps once, their total is 40 steps, but when Huck jumps four times and Tom jumps twice, their total is 58 steps.
> **Team #2** (Smiley and Grumpy): When Smiley jumps three times and Grumpy jumps twice, their total is 48 steps, but when

Smiley jumps four times and Grumpy jumps twice, their total is 56 steps.

Team #3 (Hopper and Skipper): First Hopper takes three jumps and lands in the same place as Skipper does when he takes four jumps. Then Hopper takes six jumps and nine steps to land in the same place as Skipper does when he takes nine jumps.

Your job is to figure out which pair is the winner. When you combine the lengths of one jump for each frog in a pair, which pair has the greatest length?

Before you read on, draw number lines to represent the tracks and solve each of the problems. Did the context make it easier for you? Were you able to draw the pictures of the jumps in ways that helped you? The problem context defines the winners of the pairs competition as being the two frogs whose combined single-jump length is largest. This focuses attention on a set of relationships (rather than individual values) that are critical to solving the systems of equations. This is not an easy transition. For example, in symbolizing the case of Huck and Tom, we have $3h + t = 40$ and $4h + 2t = 58$, where h is the length of Huck's jump and t is the length of Tom's jump; the objective is to find $h + t$.

Alfonso and Marcy are a bit stuck at first; analyzing the first team, they focus on finding individual jump values. Bill asks, "Why don't you think about how much further Huck and Tom traveled the second time, when they had more jumps? What can you learn just from that?" Their resulting representation of the problem is shown in Fig 8.12. They don't need to find the values of h and t separately—they find $h + t$ directly as the difference between $4h + 2t$ and $3h + t$.

Analyzing Smiley and Grumpy's sequences, students notice that Smiley has an extra jump in one of them, so they are able to determine the length of Smiley's jump readily. From there they can find Grumpy's length.

In analyzing Hopper and Skipper's sequences, the students have to work with a proportional relationship. Since three of Hopper's jumps are the same as four of Skipper's jumps, they realize that six of Hopper's jumps are the same as eight of Skipper's jumps. The nine extra steps Hopper has to take to match Skipper's jumps means that Skipper has nine steps in his jump.

In each case the students are working with relationships between unknown quantities—not only the values.

Bill structures the math congress so that relationships are the focus. The relationships he wants students to discuss are embedded in the number of jumps in these problems. When students are given random systems to work with, these strategies don't evolve. Here, however, they are on their way to making sense of more sophisticated processes for solving systems of equations and more generalizable, formal strategies can be introduced with meaning and with powerful representations.

If they haven't already begun doing so, students may begin using letters to symbolize the variables instead of representing lengths on a number line. (This, of course, is what they will do in some later algebra course.) In using this strategy, students need to distinguish between the number

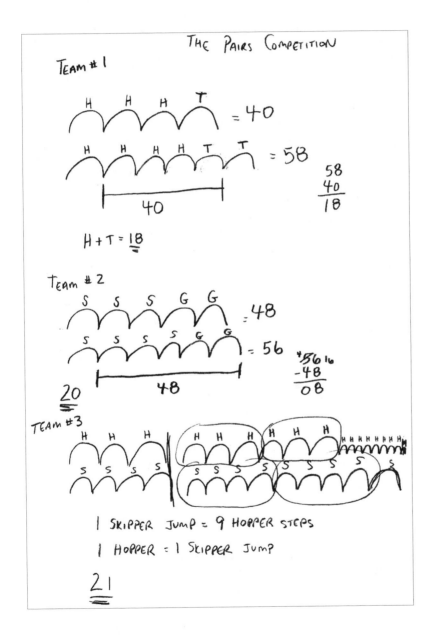

FIGURE 8.12
Student Work on the Pairs Competition

of jumps and the lengths of jumps. For example, if they represent the
first problem as $3h + t = 40$ and $4h + 2t = 58$, then it is important for
them to discuss what h and t represent. Here h represents the length of
Huck's jump and t represents the length of Tom's jump—not the number
of jumps.

The distinction between numbers of jumps and sizes of jumps is
important. Students will frequently interpret a variable as a noun instead
of as a quantity. For example, students are tempted to express the relationship that there are five nickels in a quarter as $5n = q$, reading n as
nickels and q as quarters. In the work in Cynthia Lowry's class detailed
in Chapter 5, this equivalence is represented as 5 ⑤ = ㉕ (the 5 and 25
are circled as if they are coins) to represent a constant value of a coin.
This is deliberate—we don't want variables to represent fixed known
values. When variables are used in this context of nickels and quarters,
the usual meaning will be n is the number of nickels and q is the number
of quarters. What does the equation $5n = q$ mean in this setting? Suppose
you have ten nickels, say $n = 10$. Then substituting $n = 10$ gives $5 \times 10 = q$,
which means we have 50 quarters. Do ten nickels and fifty quarters have
the same value? Of course not! What one really wants is the equation $n = 5q$,
because the number of nickels is equal to 5 times the number of quarters. If you then have $n = 10$ (ten nickels), you must have $q = 2$ (two
quarters) for equivalence, which is expressed by $10 = 5 \times 2$ (substituting
10 for n in $n = 5q$). This confounding of a noun with a quantity and
using a variable as if it is a noun is the *reversal error* (NRC 2001). Comparing-
quantities problems involving distances are used to develop a proper
understanding of the use of variables as quantities, possibly unknown,
or otherwise taking on varying values for which certain relations may
hold.

THE COMBINATION CHART AND
SYSTEMS OF EQUATIONS

Combination charts are useful tools for considering systems of equations as
well as combinations of given values. Recall the system of equations with
which this chapter begins:

$$3x + y = 40$$

$$4x + 2y = 58$$

This system works well with a comparing-quantities context (such as the
suckers and gum balls) because the increment $x + y$ passing from the
first equation to the second is apparent (it is 18). A combination chart
can represent the relations in the original equation as well as the relation
$x + y = 18$.

Look at the combination chart in Figure 8.13. Given the context, the meaning of this chart is clear. The first question in the original context asks about the movement in the chart from 40 to 58, a diagonal jump of up one, right one. All diagonal jumps of the same direction and length lead to the same increment because each is the result of buying one additional sucker and one additional gum ball. The second question asks about the reverse direction, going from the 40 to the question mark (down one, left one). Working with these patterns enables one to use the chart to solve for unknowns, and the reasons they work can be developed within a comparing-quantities context, as they were in the fence problem.

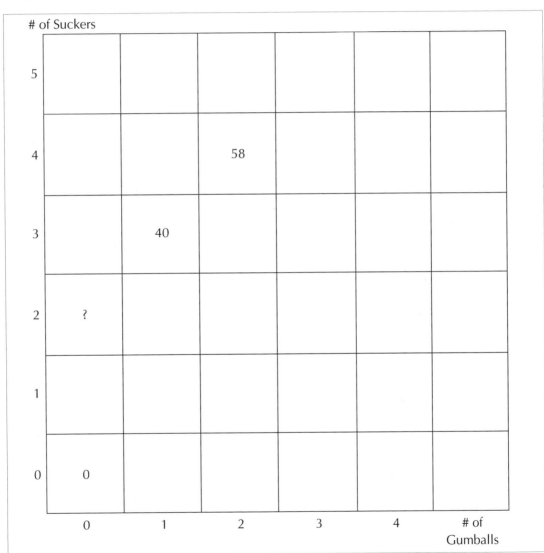

FIGURE 8.13 *Sucker and Gumball Problem on a Combination Chart*

The combination chart is a powerful representational tool that can represent many strategies for solving systems of two equations as long as the coefficients and values are whole numbers. For example, to solve

$$2x + 3y = 35$$

$$4x + 2y = 34$$

consider the combination chart in Figure 8.14. Try a left two, up one strategy to find the solution! (This will allow you to find the value of 4y.)

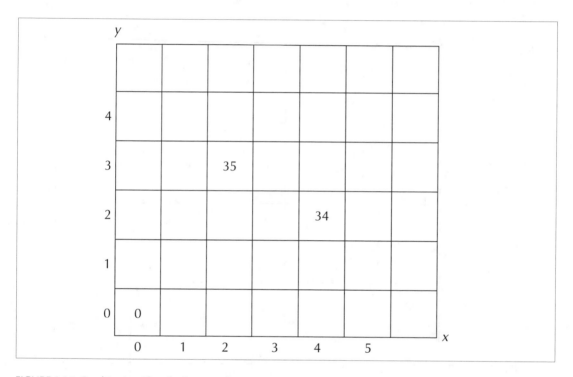

FIGURE 8.14 *Combination Chart for Systems of Equations*

The *Mathematics in Context* unit *Comparing Quantities* pushes students to work directly with combination charts to find multiple relations between pairs of unknown quantities (Van Reeuwijk 1995). In doing so they develop many of the basic operations used in linear algebra, which because of the procedure-driven approaches for solving systems in high school are often not made explicit until a college course on linear algebra. The roles of context and representation are powerful indeed!

SUMMING UP

Systems of equations in early algebra can be introduced by comparing combinations of quantities and examining relations between these combinations. If an understanding of these relationships is grounded in a meaningful context, strategies to solve for unknowns within them become easier to develop and generalize. Comparing quantities pushes students to think about relations instead of equations and procedures. The most important mathematical ideas arise through investigations of good problems. The combination chart is a powerful representation that encodes many relationships between quantities; children can describe movements on the chart as their rules that become problem-solving tools.

Variables play a crucial role in mathematics, but they represent quantities, not nouns. We have chosen contexts and representations consistent with this role, with the hope of minimizing the reversal error. However, it is *not* our goal to push symbolizing too early—students will symbolize with variables when they are ready. In the meantime double number lines and combination charts can represent the important strategies in a variety of ways. Expect a diversity of approaches in your class, and celebrate them all!

9 | DEVELOPING ALGEBRAIC STRATEGIES WITH MINILESSONS

In questions of science, the authority of a thousand is
not worth the humble reasoning of a single individual.
—Galileo Galilei

Minilessons are presented at the start of math workshop and last for ten or fifteen minutes. In contrast to investigations like the ones described in previous chapters, which characterize the heart of the math workshop, minilessons are more guided and more explicit. They include computation problems that when placed together are likely to generate discussion on certain strategies or big ideas that are landmarks on the landscape of learning. We call these groups of problems *strings* because they are a tightly structured series (a string) of problems that are related in a way that supports the development of numeracy and algebra.

Minilessons are usually presented to the whole class, although many teachers use them with small groups of students as well as a way to differentiate. During whole-group sessions at the start of math workshop, young children often sit on a rug. Older students can sit on benches placed in a U. Clustering students together like this near a chalkboard or whiteboard facilitates pair talk and allows you to post the problems and the strategies used to solve them.

The problems are presented one at a time and learners determine an answer and share the strategy they used. The emphasis is on the development of mental math strategies. Learners don't have to solve the problems *in* their head, but it is important for them to do the problem *with* their head! In other words, they are encouraged to examine the numbers in the problem and think about clever, efficient ways to solve it. The relationships between the problems in the string support them in doing this. The strategies that students offer are represented on an open number line.

Carlos, a fifth-grade teacher in California, is presenting a minilesson on variation using the following math string:

> Here is an unknown amount on a number line. I call it j.
> If this is one jump, what does $3j$ look like?
> How about one jump and seven steps?
> Now, what do three jumps and one step backward look like?
> What if $j + 7 = 3j - 1$?
> What if $j + 11 = 3j - 1$?
> What if $j + 11 = 3j - 5$?

He reminds the students about representing distances using an open number line and begins the string by drawing a small jump on the line, telling the students it is j. Then class members represent the second, third, and fourth jump descriptions. (Heidi and Alyssa's representations are shown in Figure 9.1.)

Next Carlos asks whether the drawings include an accurate representation of the equation $j + 7 = 3j - 1$. Alyssa says, "Ours isn't going to work. We're going to have to draw it all over again."

Carlos grins. "Did you draw it wrong? I thought we agreed you were correct."

"You tricked me," Heidi declares. "The equation makes what I was doing look wrong. But we didn't know then that j plus seven had to equal three js minus one."

"The problem isn't with Heidi's drawing," Alyssa adds, "we're just going to have to draw it over again."

The string has succeeded in generating a spirited conversation on variation. As pointed out in Chapter 6, variables describe relationships. All the drawings thus far are correct given the information known at the time. As more information is given—such as 3 jumps and a step back is equivalent to a jump and 7 steps—the value for j is forced. Understanding variation is a big idea on the landscape of learning. To support his learners in constructing this idea, Carlos encourages further reflection.

FIGURE 9.1
*Heidi and Alyssa's
Representations*

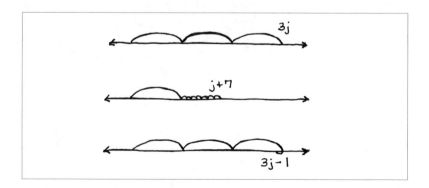

"Does everyone understand what Alyssa and Heidi are worried about?" Carlos asks. "Talk with your neighbor for a few minutes to make sure you understand the issue."

After a few minutes Carlos asks Juan to explain. "Well, the point is," Juan says, "that Heidi didn't know how big the jumps would end up. So they have to fix it. That's what Alyssa means. What Heidi did is okay, but the step length has to be related to the jump."

"Well then, what could the length of the jump be?" Carlos asks. "Heidi, do you want to add something?"

Heidi explains, "We didn't know this information before. I think the way we drew $j + 7$ is really just a way of showing there is one jump and seven steps and it's just a way to think about it. So what we did is still fine. It's just that when we know that j plus seven has to equal $3j$ minus one, then, well, like Alyssa says, we have to redraw it."

Carlos says, "Do all of you agree? Heidi is saying that this picture represents the total distance traveled in one jump and seven steps and that the amount traveled depends on the size of the jump in relation to the steps. So we can think of a whole bunch of possibilities and this picture represents just one of these possibilities." Most students nod in agreement. Carlos continues, "And the same for these three jumps minus one. It's a picture that helps us think about the possibilities, but to solve a particular question we may have to work a little more to find the jump length."

"Yeah, it's like an all-in-one picture," Juan blurts out. *Variation is now the focus.*

WHAT IS REVEALED

Carlos is using this string to encourage students to represent related algebraic expressions and to treat variables with variation. He represents j, and the students quickly agree that $3j$ is just 3 times as big, and that $j + 7$ would just be 7 small steps more. The next representation is also easy: three equal jumps. The fourth requires just one step back. The third and fourth representations do prompt students to comment that they don't know how big the jump is in relation to the steps, which is true. Nevertheless, asking students to draw what the question indicates produces a variety of meaningful representations. Even though these representations cannot be used to determine the value of a variable, the students are treating the expressions as objects. The equation introduces an equivalence that causes students to have to redraw. As the string continues, new equivalents are introduced, causing further adjustments.

Too often students believe there is only one way to complete a mathematics problem. Flexibility in thinking is crucial for making sense of the big idea of variation. When more information is provided—in this case, $3j - 1$ is equal to $j + 7$—an exact relationship between jump lengths and step lengths can be nailed down. Without this added information, there are many possible representations for the expressions $3j - 1$ and $j + 7$.

In previous years, Carlos asked his students to find mystery numbers: "Find for me the mystery number that satisfies the rule that $m + 7$ is the same as $3m - 1$." His students would set to work with calculations, trying many numbers, 1, 7, 3, 10, 5, and in time they found that 4 was the mystery number. For them, the question was based on simple arithmetic, and they knew that with a little patience they could solve the mystery. (Often his students would base a new guess on information from the previous guess, so the process was not totally random—more a guess-and-revise strategy.) Once a mystery was solved, they called out their answers and were ready for a new mystery. Students were engaged, and it was fun, but something was missing. The students were not using the relationships encoded in the mystery—they weren't structuring.

The jump contexts, number lines, and strings Carlos uses now prompt his students to think, evaluate, and reevaluate. In another four years when they are in high school, Carlos' students will study functions; the work he is doing now is an important part of their preparation. When Juan says, "It's like an all-in-one picture," he is explaining that by representing $j + 7$ and $3j - 1$ as Heidi did in Figure 9.1, a relationship is at play; for a possible jump length, there is a resulting total length indicated by the representation.[1] As the students explore this string, they use these relationships, not guess-and-check arithmetic, to think about $j + 7 = 3j - 1$.

BACK TO THE CLASSROOM

Carlos asks his students to work with a partner and draw a picture that represents $3j - 1 = j + 7$. After a few minutes he asks Sam and Ramiro to share their thinking.

Ramiro begins. "You see, we have to fit the eight steps into two jumps so we think it is four." Carlos asks for more details and Ramiro rephrases. "See, the three jumps and the backward step. Those two jumps and the backward step have to be the same as seven steps, so we have to fit seven steps, I mean, eight steps, all into two jumps. That means there are four steps in a jump."

Carlos creates the representation in Figure 9.2 and asks, "Do you want to add something Sam?"

"Yeah, it's like we used a storage box. We put the first jump both times in the storage box and just looked at the other jumps and steps."

Carlos continues to probe. "This new diagram is different from what we had when Heidi had me draw three jumps and one step backward. Is

---

[1] In terms of function language, the possible jump lengths are the input values (or domain) of the function, and the total length represents the output values (or range) of the function.

Heidi's diagram right? Give me a thumbs-up if you think her diagram is still okay." Some thumbs go up, but not too many. "Maria, you've got your thumb down, why?"

"Because it won't work. The eight steps don't fit."

"Heidi, you don't seem to agree."

"No, mine is still right. Like Juan said, mine just shows how it goes and we didn't know the amount of the jump. So it's still good. But the four steps in a jump work in Ramiro's picture because he had another equation."

"Not everyone seems convinced. Let's come back to this in a moment after we try one more problem in this string." *Variation is a difficult idea and Carlos decides to continue to examine this idea.* "Take a minute and give me a thumbs-up when you know what would happen if $j + 11 = 3j - 1$."

After a few minutes, Carlos calls on Maria. "We had to change it again. Now we had to put twelve steps into two jumps instead of eight. Now the jumps are bigger."

Rosie adds, "They are six now, because six and six is twelve."

"I agree," Heidi says tentatively, "but it's the same strategy as in my picture and that picture is still right too."

"And they are still using the storage box for the first jump like Sam said," Keisha points out. "See, the first jump didn't matter, it's the eleven plus one that matters for two jumps."

Carlos lets his students negotiate this terrain for a bit. Once they seem comfortable with Maria's answer, he asks them to consider $j + 11 = 3j - 5$. As the string continues, the class has more opportunities to consider the idea of variation, and they have a chance to talk about generalizing the approaches used for this type of problem.

WHAT IS REVEALED

In this brief minilesson, Carlos chooses a series of related problems and asks his students to solve them. Together they discuss and compare different strategies and ideas and explore relationships between problems. The relationships between the problems are the critical element of the string. As he works through it, Carlos uses the double open number line

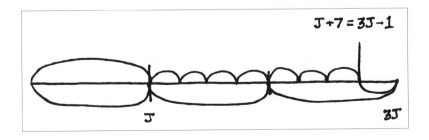

FIGURE 9.2
*Representation of
Ramiro's Strategy*

as a representational tool. This representation enables children to examine the equivalence of the expressions.

Carlos succeeds in bringing to the fore some important big ideas—that variables describe relationships and are not merely unknown quantities (variation) and that equivalent amounts can be separated off in solving a problem. The class has also had an opportunity to generalize about a strategy that is useful in considering similar problems of this type. The conversation early on with Heidi and the validation of her original representation shows students they may often have to redraw to account for variation.

Good minilessons always focus on problems that are likely to develop certain strategies or big ideas that are landmarks on the landscape of learning. Designing such strings and other minilessons to develop algebraic ideas and strategies requires a deep understanding of their development—the choice of numbers, representational models, and contexts used are not random.

JOURNEYING THE LANDSCAPE

Choosing the Numbers

Equivalence is a big idea on the algebra landscape. Underlying this idea is another big idea—that expressions can be treated as objects (rather than simply as procedures) and placed in relation to one another. These ideas are precursors to the idea that variables describe relationships and are not merely unknown quantities (variation). All of these ideas enable the important algebraic strategy of separating off equivalent amounts. Knowing that these ideas and strategies are important, Carlos has carefully crafted a string of problems to support their development along the landscape of learning.

Here's a similar string crafted around the same big ideas:

Here is an unknown amount on a number line. I call it j. Where is 2 times j?
Where is 2 times these 2 js?
Where is $4j + 6$?
Where is $4j - 6$?
Suppose I tell you that $4j - 6$ is the same as $2j$. What now?
How about $2j - 3$? Where is it? Why?
What if I told you that $4j - 6$ is the same as $3j$. What now?
Now where is $2j - 3$? Why?

In the first problem, j is represented and $2j$ would just be twice it. The next problem will probably also be easy for the students: There are four equal jumps. The third problem introduces +6 and the fourth –6 and here students may begin to comment that they don't know how big the jump is in relation to the steps. This is true, and such comments should be encouraged, but it is important to ask students to draw what the question indicates and share a variety of representations. Their diverse representations will promote discussion of this unknown quantity, since the size of the jump could vary—and now variation is up for discussion. The next problem introduces an equivalent expression that may be a surprise and most likely will cause students to have to redraw. As the string continues, new equivalents are introduced, causing further adjustments.

JOURNEYING THE LANDSCAPE

The double number line that Carlos has chosen as a representational model prompts students to examine how the expressions are related and to use equivalence. In each of the strings there are equations in which the numbers and the representation of them on double number lines potentially lead students to think about removing equivalent expressions.

In algebra classes, students traditionally are taught to "cancel out" or "add equal amounts to both sides of an equation." But too often they are taught these rules before they have made sense of what an expression such as $j + 7$ actually means. Carlos is very careful to validate Alyssa and Heidi's diagrams for $j + 7$ and $3j - 1$, because both are correct and complete. Up to this point Carlos has been working to ensure that his students understand that an algebraic expression can be treated as an object (not only as a procedure)—that the multiple representations on the number line as students suggest small steps or big steps provide possible mental images of this object.

With the added condition that $j + 7 = 3j - 1$, the diagrams have to be redrawn because the jump length is now specified by the relationship of these two expressions. In the new drawing (Figure 9.2), the representations of $j + 7$ and $3j - 1$ still have the same structure as Heidi's drawings did in Figure 9.1, but now they are aligned to end at the same point. The act of redrawing is an action in which *variation* is implicit. There is a continuum of possible diagrams for $j + 7$ that are correct, in which the length of j can stretch or shrink, and this idea is used to create a diagram in which the ends line up. It is at this point that it becomes apparent that the initial jump in each sequence can be separated off. This is what Ramiro accomplishes when he says "eight steps all into two jumps." By putting the first jump aside (like in a "storage box") Ramiro has reformulated solving $j + 7 = 3j - 1$ as equivalent equations: $7 = 2j - 1$, or $8 = 2j$. He has used equivalence to remove an instance of the variable j.

Choosing the Model

The double number line is a powerful tool for examining equivalence and developing algebraic strategies in the early grades, too. Patricia is using the following string with her second graders to help them treat numeric expressions as objects:

$$10 = 5 + 5$$
$$10 + 10 = 5 + 5 + 5 + 5$$
$$5 + 20 = 10 + 10$$
$$5 + 20 + 4 = 4 + 10 + 15$$
$$13 + 8 + 6 = 5 + 9 + 13$$

Each time she writes a statement, she asks the children to determine whether it is true or false. If the statement is false (the third equation in the string, for example), it is made true by replacing the equals sign with an inequality sign.

The string begins with an easy equation that supports the second, third, and fourth. The hope is that as the children work through the string, someone will suggest that a determination can be made without adding up all the numbers—that the numeric expressions can be treated as equivalent objects. For example, in the last problem the students who see $8 + 6$ as an object equivalent to $5 + 9$ won't have to add the numbers to see whether the statement is true.

The class is discussing the fourth problem in the string. Patricia writes the statement $5 + 20 + 4 = 4 + 10 + 15$, and asks her students to give a quiet thumbs-up when they are ready to say whether the statement is true or false. (Her students have learned not to disturb their classmates by calling out answers. Also, thumbs held up in front of the chest allow class members who might be distracted by waving arms crucial time in which to think.)

When most thumbs are up, Patricia calls on Ian. "I say it's true," he declares with conviction.

"Okay, it looks like others agree, so tell us why you think that."

"Well, the fours don't matter because they are on both sides. And the five plus twenty is a twenty-five and the ten plus fifteen is another twenty-five. So it's true."

For many children, the only way to solve this problem is to use an arithmetic strategy: Add up both sides and compare answers to see if they are the same. For example, here they would produce $29 = 29$. But Ian's first sentence shows this isn't his strategy—he is using a more algebraic strategy by noticing the equivalence of the 4s on each side of the equation and ignoring them. He is also implicitly using the commutative property of addition ($25 + 4 = 4 + 25$)—it doesn't matter the order in which you add things.

Patricia pushes Ian to say more.

"Because you are adding them up and four more is the same as four more," Ian clarifies.

"How many of you can explain what Ian is thinking?"

Mia gives it a go. "I think what Ian is saying is that five plus twenty is twenty-five and that ten plus fifteen is twenty-five, and these are the same, so when you go four more you get the same thing and so it's the same on both sides."

Patricia draws the representation shown in Figure 9.3 and asks Ian and Mia if this represents their thinking. Mia Chiara nods in agreement, but Ian is not convinced.

"Ian, you're not convinced?" Patricia says. "But you were sure a moment ago."

"No, I'm convinced the equation is true, but your picture isn't what I'm talking about."

"What's not right about the picture?"

"Well, I said the fours don't matter, but you put them both on the same side and that's not where they are." Patricia had placed the fours one above the other on the number line, hoping to show the equivalence Ian had mentioned. Surprisingly, although Ian has used equivalence in his determination that the statement is true, when representing addition on an open number line the order matters to him. Patricia draws another double number line on which the jumps appear in the order presented in the equation (see Figure 9.4) and asks Ian if this is what he means.

"Yes, but you have to put the twenty-fives in."

Patricia adds the 25s (see Figure 9.5) and Ian nods.

Sensing that the big idea of the commutative property needs further discussion, Patricia asks the children to talk with their partner about these two diagrams and about how Ian and Mia are thinking about the problem.

Rosie offers some thoughts, "I think they both see the twenty-fives, but Ian wants to add the fours first and second while Mia wants to add the fours at the end both times. But it doesn't matter, really."

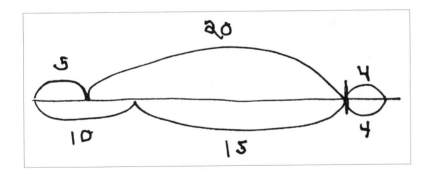

FIGURE 9.3
*First Representation
of Ian and Mia Chiara's
Strategy*

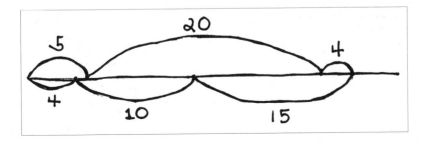

FIGURE 9.4
*Second Representation of
Ian and Mia Chiara's
Strategy*

"Yeah," adds Camille, "because you are just adding them up and you can do it both ways."

"That's what I meant when I said the fours don't matter because we are adding it all up," Ian points out. "They don't matter. Both are four and twenty-five." *The landmarks of commutativity and equivalence have now been reached.*

CHOOSING A CONTEXT

Teaching and Learning in Another Classroom

Maia's second graders are used to playing a version of twenty questions in which the goal is to determine what coins (nickels, dimes, pennies, and quarters totaling fifty cents) she has in her hand. Today she is playing a variation on this game in her minilesson. She has set down two small bags with coins in them; the children are to determine which bag has the most money or whether they contain equal amounts. The first bag has one quarter, three dimes, four nickels, and one foreign coin; the second bag has one quarter, two dimes, six nickels, and one foreign coin. Both foreign coins are the same but of unknown value. The children are midway through the game.

"So far you have figured out that there is one quarter in each bag and two dimes in this bag on the left. Let me write that down. What sign should I use so far?" asks Maia.

"Greater than" several children call out.

"Why, Sam?"

"Well, the quarters are the same, but the bag on the left also has two dimes." Sam has used equivalence to separate off equal amounts. He doesn't need to add the known values in the bags, because the quarters are equivalent.

FIGURE 9.5
Adding in the 25s

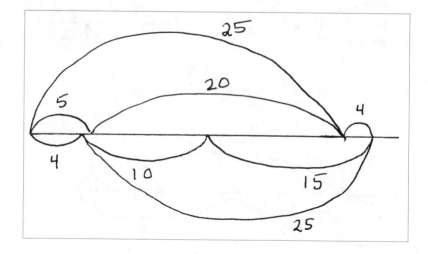

Maia writes $1\,\widehat{25} + 2\,\widehat{10} > 1\,\widehat{25}$ on the board. "Okay. Do you have more questions for me?"

Juanita asks, "Are there more than two dimes in the other bag?"

"Yes," replies Maia.

Kelly gets specific. "Are there three?"

"Do you mean exactly three?" Kelly nods. "Yes. So let's write this down. Now you know the quarters and the dimes. Which sign do we need?"

"You have to turn the sign around," replies Juanita.

Maia writes $1\,\widehat{25} + 2\,\widehat{10} < 1\,\widehat{25} + 3\,\widehat{10}$ on the board.

After a bit more discussion, the numbers of nickels are determined, pennies are ruled out, and the statement $1\,\widehat{25} + 2\,\widehat{10} + 6\,\widehat{5} = 1\,\widehat{25} + 3\,\widehat{10} + 4\,\widehat{5}$ is established. Explaining this Kelly says, "There is one more dime and two less nickels in the second bag, so they're the same because one dime equals two nickels." *Rather than adding all the values up to prove the values in the bags are equal—an arithmetic strategy—Kelly is using an algebraic strategy. She is mentally substituting a dime in the right-hand bag for two nickels.*

Now Maia teases the children with a smile. "You think you're done now, don't you? Actually there is one more coin in each bag and it is the same type of coin in each bag."

"What is it?" Juanita asks what everyone is wondering.

"I don't know what this type of coin is worth. They are foreign coins that I got a long time ago when I was traveling. I put one in each bag. Let's call it *c* for coin because we don't know what it is worth. What sign should we use?" As the children ponder her question she says, "Here is what we know so far," and writes $1\,\widehat{25} + 2\,\widehat{10} + 6\,\widehat{5} + \widehat{c} \,?\, 1\,\widehat{25} + 3\,\widehat{10} + 4\,\widehat{5} + \widehat{c}$.

"We can't do it if you don't tell us," Sam says with exasperation. "How can we add it if we don't know what it is?"

Sam and his classmates are still relying on arithmetic to verify equivalence, so the problem seems impossible. They do not yet have a strong sense of equivalence, and they also assume an expression represents a procedure. It is impossible to add something to something if you don't know what the something is. Maia wants to make sure her students develop an understanding that equivalence can be understood without computation.

She asks the class, "Do you have to add it?"

"If it's a nickel, it's still equal," replies Rosie.

"It works for a penny or a dime, too," adds Juanita.

"Does it work for other numbers, too?" *Maia encourages students to consider several numbers as a way of developing the idea of a variable. But here, the variable is not an unknown number that has to be found. Instead it represents many possibilities (its value is in some range of amounts)—once again this is variation.*

Keshawn begins to grasp the big idea. "It works for any number, because it's the same in both bags. You don't have to know what it is."

"Yeah, Keshawn, you're right!" Isaac says in awe. "If it's the same coin and it's in both bags you don't have to worry about it. It's the same on both sides so it's still an equal sign."

Maia continues to use the context of the foreign coin to support the development of variation. "Could c be any amount? Are Keshawn and Isaac right? Would this statement be true no matter what c is?" She writes in an equals sign: 1 ㉕ + 2 ⑩ + 6 ⑤ + ⓒ = 1 ㉕ + 3 ⑩ + 4 ⑤ + ⓒ.

Isaac replies with conviction, "Yep, *c* could be any number. As long as they are the same, it doesn't matter what the *c* coin is worth. It could be any number."

Maia has pushed her students to consider the big idea that equivalent amounts can be separated off, or substituted, even when variables are involved. By necessity her minilesson has also involved the big idea of variation. In addition, the children have had to view expressions as objects, not merely as describing a set of operations. This interwoven web of ideas is critical to the development of algebra.

Operating on Expressions

Once equivalence is well understood, children can be challenged to consider how one can operate on expressions. Carlos is using the following string to develop this idea with his fifth graders:

> Here is one jump and two steps. What else could it look like?
> So how about two jumps and four steps, how would I represent that?
> What about $2(j + 2)$?
> What about $3(j + 2)$?
> What about $4(j + 2)$?
> What about $2(2j + 4)$?

Carlos begins his string by drawing the representation shown in Figure 9.6. "Okay, here is one jump and two steps. Is this how it always looks? Carrie, what do you think?"

"Well, we usually keep the steps the same in our drawings, but the jump could be shorter or bigger, we don't know."

"So it could be like this, or this, or this?" Carlos draws three different representations of $j + 2$ (see Figure 9.7). Most of the students nod in agreement. "So how about $2j + 4$? Carrie, do you have a suggestion on this one?"

"Well, we do what you just did, except we do two jumps and four little steps. Each time the jumps have to be the same because of the frog

FIGURE 9.6
*One Jump and
Two Steps*

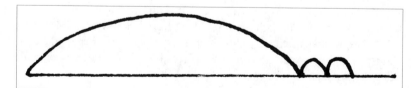

jumping rule." Carlos invites her to draw two jumps and four steps
below his first drawing of one jump and two steps on a double number
line (see Figure 9.8). "What do you think? Jasmine, do you have something to add?"

"Wouldn't it have been easier just to add a jump and two steps
rather than starting all over again?" Jasmine offers shyly. Jasmine adds
her idea to Carrie's drawing (see Figure 9.9). Next Carlos asks the students to talk with their partner about what Jasmine and Carrie have
drawn.

"I think it's two *j* plus two," Mario explains. "So no matter how you do
it, it has to be the same."

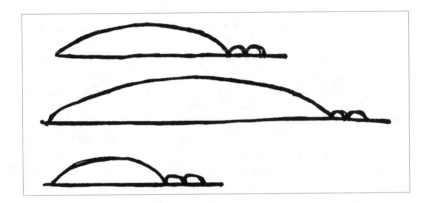

FIGURE 9.7
*Other Versions of
One Jump and
Two Steps*

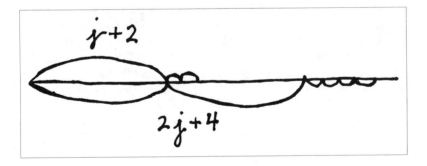

FIGURE 9.8
*Two Jumps and
Four Steps*

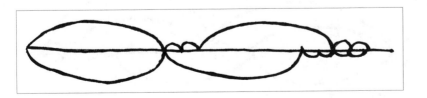

FIGURE 9.9
*One Jump and
Two Steps Twice*

Noticing that Louisa seems puzzled, Carlos attempts to bring her into the conversation. "Louisa, do you agree?"

"Well, I'm not sure. Doesn't it have to be the same *j* each time? What if they were different frogs or something?"

Mario helps clarify. "No, it has to be one, the same frog for each jump. That's what we always do."

Carlos asks Mario to add to the diagram and show everyone where the two *j* + 2's are. Mario adds two arcs over the jumps and two steps on the top of the double number line and writes 1 and 2 above each arc (see Figure 9.10).

Once again Louisa looks puzzled. "Okay, I see them, but why did you write 1 and 2 above them? They aren't one jump or one step."

"Well, this is one jump plus two and this is the other jump plus two, so I just labeled them 1 and 2." *Mario is treating the j + 2 as a single object and is operating with that expression as an object. This is a landmark in development analogous to the leap young children take when they begin to unitize a group of ten objects and count it as one ten. The ability to look at mathematical expressions and pull out chunks that can be viewed as single "entities" is an essential strategy throughout mathematics (collegiate and beyond!).*

Although Carlos has used the letter j in earlier discussions with his students, he has refrained from symbolizing to make sure that students are first making sense of the context and the representation. Now he senses the class is ready to interpret the symbolic expression, and he wants them to see correctly formulated algebraic equations that capture the focus of the minilesson.

Pleased with the discussion, Carlos asks the class, "Can we do this?" and writes $2(j + 2) = 2j + 4$ above Mario's work. The class nods.

Mario interjects, "Yeah, that's just writing what I'm saying another way."

Because Carlos knows that many beginning algebra students will read $2(j + 2)$ as $2j + 2$, interpreting the symbols left to right, rather than seeing the $j + 2$ in the parentheses as an object, he continues with the string, having his students work with $3j + 6$ and $4j + 8$. Again they work with multiple chunks of $j + 2$ as they discuss how these jumping sequences are related. As the string progresses, equivalences like $3(j + 2) = 3j + 6$, $4(j + 2) = 4j + 8$, and $2(2j + 4) = 4j + 8$ are represented on double number lines and discussed.

FIGURE 9.10
Two Times One Jump and Two Steps

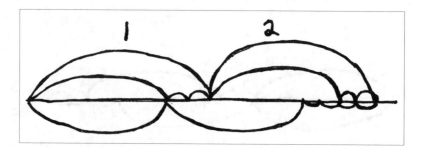

Good minilessons always focus on problems that are likely to develop certain strategies or big ideas that are landmarks on the landscape of learning. The big ideas of variation, equivalence, expressions as objects, that variables describe relationships and strategies such as using cancellation, commutativity, or equivalence can be further developed in minilessons. Designing strings or other minilessons to develop these ideas requires a deep understanding of the landscape; the choice of the questions and the models and contexts used are not random.

When Galileo said, "In questions of science, the authority of a thousand is not worth the humble reasoning of a single individual," he was describing his life-long struggle to have scientific ideas accepted in an era where religious authority dominated. But we cite it here for different reasons. When the humble reasoning of children is valued and nurtured in mathematics classrooms, doors open. When children are given the chance to structure number and operation in their own way, they see themselves as mathematicians and their understanding deepens. They can make sense of algebra not as a funny set of rules that mixes up letters and numbers handed down by the authority of thousands but as a language for describing the structure and relationships they uncover.

10 | PROOF

Don't just read it; fight it! Ask your own questions,
look for your own examples, discover your own proofs.
Is the hypothesis necessary? Is the converse true?
What happens in the classical special case? What
about the degenerate cases? Where does the proof
use the hypothesis?

—Paul R. Halmos

A BIT OF HISTORY

Questioning, defending, justifying, and proving are all processes characteristic of human activity. Even young children do these things naturally. While there are rules of logic that govern a rigorous notion of mathematical proof, the idea of proof has its genesis in the need to establish certainty and to convince others (Stylianou, Blanton, and Knuth 2009).

Professional mathematicians acquire their understanding of proof by participating in their community. For them it involves rigorous reasoning, without gaps, that establishes the validity of a mathematical statement based on clearly formulated assumptions. Although courses on proof have begun to appear in the undergraduate curriculum in recent decades, mathematicians have for the most part developed their conception of proof in settings in which their early attempts were questioned: "How did you get from here to here?" or "I believe this, but why is your next statement true?" These questions trigger an act of reflection, a reorganization of ideas, and the presentation of reasons that "fill the gaps." Mathematicians relive this process throughout their careers and publish their written proofs in peer-reviewed journals.

In early civilization, mathematics was bound up with practical questions of computation (commerce), geometry (surveying land), the passage of time, and the location of stars. One of the earliest mathematical manuscripts is the Rhind Papyrus, attributed to an Egyptian scribe, Ahmes, and written approximately 1600 B.C.E. It describes how to multiply numbers using successive doubling and explains how to find the areas of triangles and approximate the volume of cylinders. A long section describes solutions for dividing 2 by odd numbers, expressing the result as sums of unit fractions, and there is even early algebra in the text. Perhaps the most remarkable calculation of antiquity was made by the Greek mathematician Eratosthenes (266–195 B.C.E.). (He was also the third librarian of Alexandria.) Using the elevation of the sun at solstice in Alexandria and Syene, he calculated the circumference of Earth within 20 percent of its actual length.

Yes! Eratosthenes knew Earth was a sphere, and he even calculated the angle of the tilt of Earth's axis.

Eventually an interest in mathematical relationships, particularly in geometry, so captivated philosophers and mathematicians that they had to develop language in order to communicate these new ideas. Eudoxus of Cnidus (408–355 B.C.E.) was a student of Plato. A mathematician and astronomer who investigated proportion, Eudoxus expressed the idea that the area of a circle is proportional to the diameter squared: "Circles are to each other as are squares on their diameters." The area of a circle of radius r is $A = \pi r^2$, and since the radius is half the diameter d, $A = (\pi/4)d^2$. So the area of a circle is proportional to its diameter squared, the proportionality constant being $\pi/4$ (approximately 0.7853).

Although Eudoxus' formulation is opaque to us today (he didn't mention $\pi/4$), it illustrates the emergence of the language necessary to describe mathematical relationships with precision. In fact, the development of mathematical language has been a long and torturous process, especially in algebra. Our modern notation using variables, as well as the order of operations taught today, took several thousand years to evolve from descriptions like that of Eudoxus. It is no surprise that learners need time to make sense of today's algebraic expressions. And, developing that understanding does to some extent follow the historical development (van Ameron 2002).

The modern notion of *deductive proof* is attributed to Euclid (active in Alexandria during the third century B.C.E.), whose *Elements* derived the plane geometry studied today from five basic assumptions, or *axioms*. (An axiom, or *postulate*, is a statement that is not proved but considered obvious or self-evident or is otherwise taken as a basis for further work.) Although modern refinements have cleared up some of the details, the accomplishment of Euclid was monumental: He showed how basic assumptions, together with careful definitions and language, could provide the starting point for deriving all the main results of plane geometry. Euclid built proofs with a chain of reasoning, step by step, starting with definitions and axioms. This is called a deductive approach. For two millennia much of geometry teaching has been based on Euclid's work.

Not all proofs during this period were deductive, however. Perhaps the most spectacular proof of ancient times was the determination by Archimedes of the volume of a sphere. He proved that if a sphere is compared with a cylinder of the same radius and a height twice the radius (see Figure 10.1a), then the ratio of the volume of the cylinder to that of the sphere is 3:2.[1] Archimedes did not give a deductive proof, but instead relied on a physical model. He imagined a sphere, a cylinder, and a cone

---

[1]To see this, the volume of the cylinder is base area (πr^2) times height ($2r$), which is $2\pi r^3$. The volume of a sphere is $4/3\pi r^3$, so the ratio is $2:4/3$, or 3:2. Turning this around, if you know the volume of the cylinder, then you have found the volume of the sphere multiplying by $2/3$.

hanging on a bar where the different distances could be selected so that balance was achieved (Figure 10.1b).

In addition to being a tool for thinking, models can be a tool for developing the language needed for expressing mathematical ideas and for organizing proof—paralleling the development in ancient times. Legend has it that Archimedes believed his calculation of the volume of a sphere to be his greatest accomplishment and asked that a cylinder and sphere be placed upon his tomb that would bear the inscription 3:2. Seeing that ratio, everyone would know who lay there—no need for his name!

Today's mathematicians craft proofs using a variety of techniques, such as deduction (the approach used by Euclid), induction (showing how to start and also how to increase one at a time), contradiction (showing that if something isn't true, contradictions arise), and exhaustion (checking all cases). Many reason with representations like Archimedes did. And still others use computers to examine all cases. No matter what form of proof is used, two ingredients are critical: (1) the language and definitions must be clear (no ambiguity), and (2) each statement must follow logically from information previously established using mathematically accepted *rules of inference*.

A basic rule of inference, first analyzed in abstract form by Aristotle, is *modus ponendo ponens* (usually abbreviated *modus ponens*). This rule of logic says that if we know that A implies B, and if we know A, then we may conclude B. One way to build a proof is to find a sequence of statements linked together by *modus ponens*. Many deductive proofs are built up this way, starting with basic assumptions and definitions followed by applications of *modus ponens* or other related deductive rules.

A second basic rule of inference is *universal generalization*. To find out if something is true mathematicians often begin by looking at examples. But checking examples isn't usually enough to see whether something is

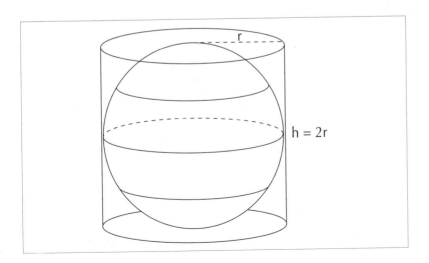

FIGURE 10.1a
Sphere in Cylinder

always true, for there may be many, even infinite cases. However, if an example checks out, and if we can examine the example thoroughly to see that no special assumptions were used in the reasoning that would limit the approach to all the cases, then universal generalization allows mathematicians to claim the result *for all*.

CHILDREN AND PROOF: WRITING IN MATHEMATICS VERSUS DEVELOPMENT OF PROOF

The notion of proof evolved throughout the history of mathematics, typically with increasing rigor but with surprises, too. Children's ideas of proof also develop over time if we prompt them to ask why and to develop con-

FIGURE 10.1b
Archimedes' Balance of Cone, Cylinder, and Sphere of Radius r

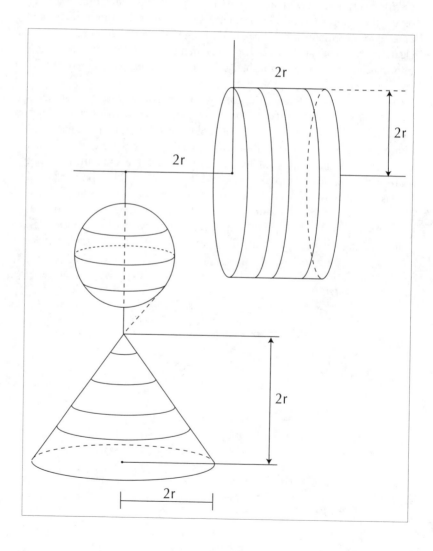

vincing arguments. If we take seriously the idea of treating young children as emerging mathematicians, initiating them into a community of mathematical discourse around proof is critical. Rather than beginning with instructions to "use words, pictures, and symbols to show your work," children need time to write and revise arguments for their peers to consider and to read and comment on one another's arguments. The development of increasingly precise language while discussing mathematical objects is crucial to the early development of proof. And the use of context and models as tools for thinking may facilitate the development of language, verbal and written.

Learners' early writing about mathematics is usually a description or *retelling* of their approaches in solving a problem. This is an important step in learning to articulate one's reasoning, and in early grades it often marks the onset of an awareness of one's own reasoning. Children may write something like, "At first we did this, but it didn't work so we tried that, and then we added it all up and we found our answer." They retell the story of what they did. Although such writing might develop language and help students focus on clarity, it is only a beginning and should not be confused with proof. Providing proof to a community requires a reorganization of ideas into a chain of logically connected statements. These steps involve *analysis and resequencing of a process,* in search of a *convincing chain of reasoning,* rather than retelling what was done. To do this requires that learners take the perspective of the audience.

Children need many experiences reexamining, revising, and simplifying their ideas in order to make concise arguments for readers if they are to build a foundation for understanding a mathematician's view of proof. Teaching practices in which children are encouraged to read and question others' ideas, examine their own and others' thinking, develop conjectures and build arguments for these conjectures, or write and talk about their reasoning is foundational for building a habit of mind toward proving (Stylianou, Blanton, and Knuth 2009). When we provide an audience—readers—and ask children to write up their "proofs," shared principles and rules of deduction begin to be used in a chain, building on one another. Children naturally reason using universal generalization, and they develop systematic ways to explore all cases.

TEACHING AND LEARNING IN THE CLASSROOM

As a context for modeling the open number line (although some of her students had previously used the open number line to represent addition and subtraction strategies, many had not), Leah's fifth graders are investigating the possible two-color farm fences that have a length of 32 using Cuisinaire® rods of length 3, 4, 6, and 8. Each length is a specific color, so a "two-color fence" is built using rods in two of the lengths. The restriction to lengths 3, 4, 6, and 8 also prompted the development of equivalence and multiplicative relationships: The students need to choose two of these lengths and find the possible ways to build 32. Earlier in the week the students used the actual Cuisinaire® rods, but now Leah has put the rods away and instructed them to use relationships between 3, 4, 6, and 8 they had discovered earlier and represent their ideas on an open number line.

Harry and Margarita have been working with 3s and 6s, and they don't think it is possible to build 32. Leah asks them to record their observations in preparation for a gallery walk the next day. They draw jumps on an open number line and write on their poster:

> First we tried all 3s but that didn't work since $3 + 3 + 3 + 3 + 3 + 3 + 3 + 3 + 3 + 3 = 30$ is too little but if we do 3 more we get more than 32. So we tried one 6 and more 3s but that still didn't work. We tried using two and three 6s but that didn't work either. So it is impossible.

Harry and Margarita are retelling what they did. While they can list all possibilities and provide a full set of reasons, they have not yet considered how to show that all possibilities have been considered—a critical piece in structuring the relationships sufficiently to craft a compact proof.

Sonia and Alice start the same way, adding up 3s, but when they use 6s they realize that it is the same as adding 3s. Leah sees the girls writing addition sentences in lists and says, "Alice, you just said a six is the same as two threes. Is there a way you can use this equivalence to help you here? And look at all those threes you are adding, can you write that another way?"

"You mean like with multiplication?" Alice responds quizzically.

"Yes, see if you can rethink what you are trying to do with multiplication and write an easier and more convincing proof for your classmates to read." *Leah is pushing the girls to structure their work differently in the search of a better proof.* After a while Sonia and Alice write:

> Fences made only from 3s have to be multiples of 3, and 32 is not a multiple of 3 because 30 is a multiple of 3 and the next multiple of 3 is 33. If you use 6s, it's like using two 3s, so you still only get multiples of 3. So you can't make 32 from 3s and 6s.

Sonia and Alice have restructured their work using multiplicative instead of additive relations. But, more important, the restructuring leads to a new chain of reasoning, which in this case is more direct. Later, in the math congress, Sonia adds more: "We knew that it wasn't enough just to calculate examples because there may be other possibilities. So that's why we did it for 3s and then saw that if you used 6s you didn't get any more new numbers." *Sonia has constructed a big idea about proof, that listing examples may not be enough. This idea is also an important motivator: Sonia and Alice had to look for a different approach so they wouldn't have to give a long list of possibilities.*

Mark and Daniella have produced a list of computations involving 6s and 8s trying to find which add to 32. They have circled those that work and crossed out those that don't. Leah encourages them to review their work. "For your poster, try to craft a proof that you have them all. Can you review all that you have done and try to organize it in a way that will give a convincing proof without retelling all that you did? Maybe you can make it even easier." *Leah is pushing Mark and Daniella to look for relationships in their work and to write them down—this is very much like the revision process we encourage in young writers.* Mark and Daniella want to make sure their list is complete and want to examine all cases. But they need to analyze and resequence their process. Daniella decides they can create a complete list if they organize by counting the number of 8s. She records:

> We worked with 6s and 8s. First we did one 8 and we needed 24 more so we used four 6s. Then we did two 8s, but we can't make 16 out of 6s so we tried three 8s and that didn't work either because we can't make 8 out of 6s. Finally four 8s work without any 6s. So there are two ways, $8 + 6 + 6 + 6 + 6 = 32$ and $8 + 8 + 8 + 8 = 32$. P.S.: You can't do zero 8s either because five 6s is already 30.

Mark and Daniella have visited several important landmarks in the development of proof. They are using a list of cases and are also justifying why their list is complete—this is proof by cases.

WHAT IS REVEALED

The context of measuring fences for farms was an opportunity for Leah to make sure her students understood the concept of an open number line. The inquiries she chose motivated them to use equivalence and multiplicative structuring. The problem of two-color fences led to proofs based on examining all cases or using divisibility rules and thus provided an opportunity for these students to make an important transition. All six children began by adding numbers to see what would work and what wouldn't work. They represented their fence combinations on open number lines— one of Leah's goals. But when asked to create proofs for a gallery walk, they started by retelling their steps rather than analyzing what they had done and trying to resequence their ideas. Leah had expected this, because the previous year the teachers had students keep math journals and instructed them to write about what they did. But now Leah wants to push students toward crafting more elegant proofs by reexamining their work and writing for an audience.

Although Harry and Margarita retell the sequence of calculations they made, the choice of numbers in the context inspires Sonia and Alice to structure the problem multiplicatively, and this leads them to a new chain of reasoning based on divisibility by 3. In this case, a significantly more compact proof emerges—they discover it is impossible to use 3s and 6s only, because 32 is not a multiple of 3. Mark and Daniella know they need to use a complete list of cases and be sure to show why their list is complete. They know that listing examples may not be enough. They organize their list by the number of 8s they try, and they have a rather complete proof to share with the class. Later, in the math congress, they will hear about yet another way to think about their cases, based on the equivalence that $4 \times 6 = 3 \times 8$.

DEVELOPING A SYSTEM OF SHARED PRINCIPLES AND RULES OF DEDUCTION

Bill's fifth graders (see Chapter 6) were able to construct a cancellation rule involving variables. They constructed their own terminology for this process, describing cancellation as placing equivalent amounts in a storage box and establishing a deductive rule like the following: Given $a + b = a + c$, deduce both $a = a$ and $b = c$. In this context involving unknowns the students were unwilling to give up the three equal jumps on both sides of the equation $3j + 6 = 4j - 2$. In time they will, but at the moment their shared principle allows them to conclude $3j = 3j$ and $6 = j - 2$.

Second graders (see Chapter 5) worked with equivalence while playing the piggy bank game and analyzing true and false statements. They developed an understanding of compensation (what is added must be removed to maintain identity), the commutative and associative properties of addition, ignoring (cancelling out) like amounts, and adding and/or subtracting n to both sides to simplify for analysis. When children are motivated to convince their peers, deductive rules can emerge naturally, even at early ages, and the double number line remains a powerful model for them to reason with.

What can second graders do if we ask them to write up their proofs and provide an audience of readers? Will these shared principles and rules of deduction begin to be used in a chain, building on one another? Will their written arguments take on any early form of proof?

To examine this question, Patricia Lent (see Chapter 7) wrote up several equations on a paper, and we asked the children to write out their proofs for us to read. Several pieces of work are included in Figures 10.2a, b, and c.

Note how the children's arguments are sequential: "First I knew 4 + 5 = 9 and there was a 9 on the other side, so I crossed those out. On one side was 5 − 5, so I crossed that out. Then there was a 2 on the other side so it wasn't equal." Although they are retelling their strategies, the strategies are based on rules that have been established in the community. The arguments exemplify cancellation ("there are two Ns so they're out"), the commutative law ("I am switching minus 3 to 11 instead of 20"), and a form of compensation sometimes called swapping ("if I switch the 7 from the 17 and put in the 5 then I have 15 + 7 = 15 + 7"). The community's rules of deduction have become tools for framing coherent mathematical discussions. Although they start with retelling, because they are using deductive rules instead of describing computation, they are moving toward proof.

Even more interesting is the use of the words *if* and *then*. These two words are the ingredients of the most important structure in mathematical logic: that of logical implication. The work in Figure 10.2a says, "If 20 + 9 = 10 + 19, then if I take *n* away from one side and not the other

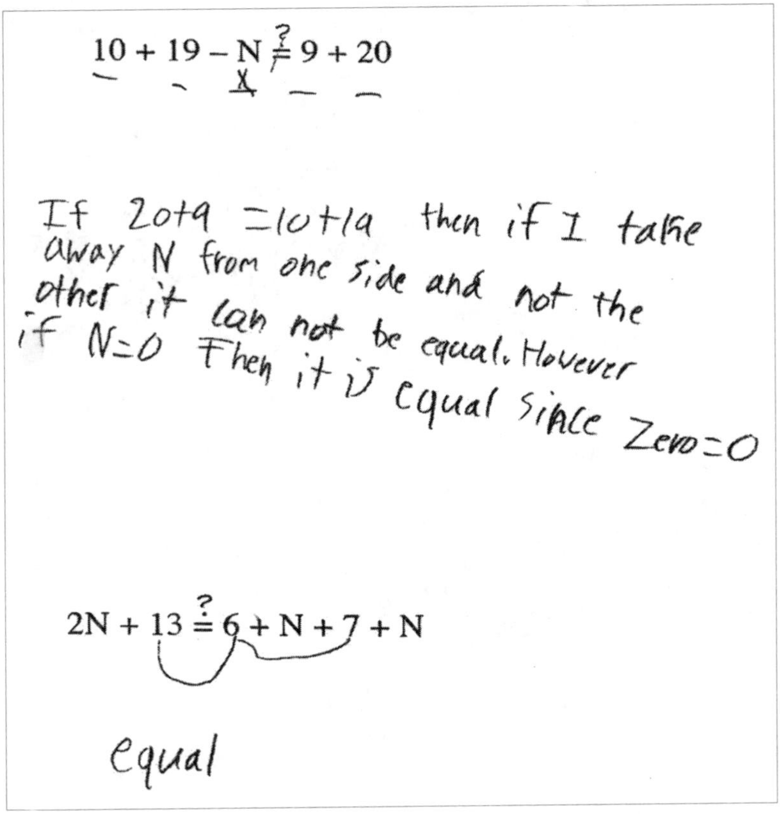

FIGURE 10.2a
Trish's Student Work, Equal or Not Equal

then it cannot be equal." Equivalence given by compensation has been accepted in their classroom; since this student believes $20 + 9 = 10 + 19$ and has also observed that if $20 + 9 = 10 + 19$ then $20 + 9 + n = 10 + 19$

FIGURE 10.2b
Patricia's Student Work, Equal or Not Equal (continued)

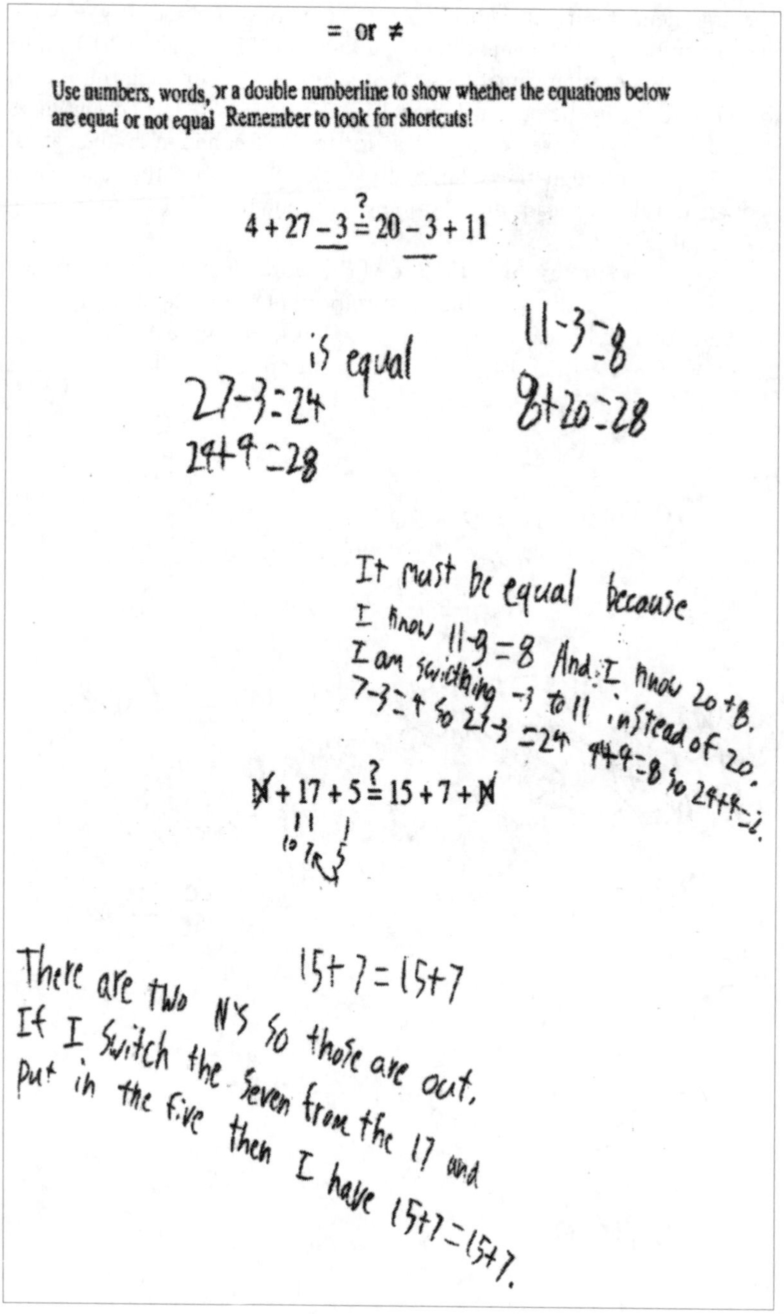

= or ≠

Use numbers, words, or a double numberline to show whether the equations below are equal or not equal. Remember to look for shortcuts!

$$4 + 27 - 3 \overset{?}{=} 20 - 3 + 11$$

is equal

$27-3=24$
$24+4=28$

$11-3=8$
$8+20=28$

It must be equal because
I know $11-3=8$ And I know $20+8$.
I am switching -3 to 11 instead of 20.
$7-3=4$ so $27-3=24$ $4+4=8$ so $24+4=8$.

$$\cancel{N} + 17 + 5 \overset{?}{=} 15 + 7 + \cancel{N}$$

There are two N's so those are out.
If I switch the seven from the 17 and
put in the five then I have $15+7=15+7$.

$15+7=15+7$

cannot be true (except if *n* = 0)—an early use of *modus ponens*, one of the most treasured forms in mathematical logic!

GALLERY WALKS, ASKING QUESTIONS, AND REVISITING THINKING

This work, as well as that of other researchers, prompted us to ask ourselves several questions. Are these chains of reasoning more than just

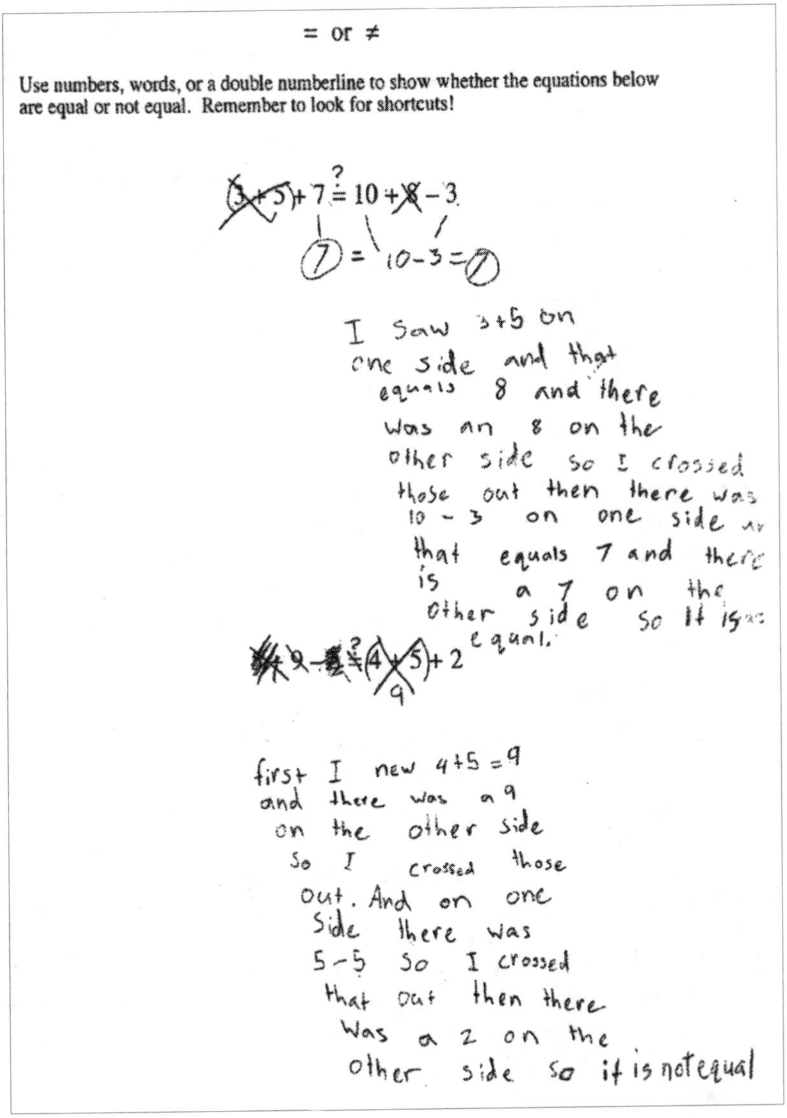

= or ≠

Use numbers, words, or a double numberline to show whether the equations below are equal or not equal. Remember to look for shortcuts!

$(3+5)+7 \overset{?}{=} 10+8-3$

$7 = 10-3 = 7$

I saw 3+5 on one side and that equals 8 and there was an 8 on the other side so I crossed those out then there was 10 - 3 on one side an that equals 7 and there is a 7 on the other side so it is equal.

$4 \times 2 - 5 \overset{?}{=} (4 \times 5)+2$

9

first I new 4+5 = 9 and there was a 9 on the other side So I crossed those Out. And on one Side there was 5-5 So I crossed that Out then there Was a 2 on the other side so if is not equal

FIGURE 10.2c
Patricia's Student Work, Equal or Not Equal (continued)

retelling or giving evidence? Can we classify them as emergent forms of proof characterized by use of deductive rules? If children continued and were supported in this type of work, how might it develop?

To answer this last question, over the past several years we have had children participate in "gallery walks," in which they write up their initial proofs, post them for everyone to see, and invite peer comments. These gallery walks take place before a math congress begins and are akin to consultations of professional mathematicians: Mathematicians regularly talk about their ideas with colleagues in order to clarify and simplify their work and also to weed out anything incorrect.

First, children write up their strategies and ideas in ways that explicitly attempt to convince readers. These "presentation" pieces are then posted around the room and students walk around with sticky notes. They read the pieces, make comments and queries on the sticky notes, and put the notes on the posters. We introduce this work by explaining that when mathematicians write up their findings for math journals, they do not merely reiterate everything they did. Instead, they focus on crafting a convincing and elegant argument, or proof, for other mathematicians to consider. Consultation and communication is an important part of the way mathematicians prepare their work and convince one another.

Of course, elementary students are not expected to write up formal proofs, but by focusing on the justification and logic of their arguments, over time they do become better able to eliminate extraneous detail and delve into the important ideas. As children walk around and comment they are encouraged to read critically, to follow the reasoning, to look for holes in the thinking, to disagree, and to question. At first their comments are quite superficial—"I think you made a really pretty poster"; "I like the way you used colors"; "I agree with your answer." But over time, aided by discussing the comments they find helpful, they begin to ask better questions—"I understand the top three statements, but I don't understand how you got from there to your answer."

BACK TO THE CLASSROOM

Let's return to Miki Jensen's classroom (see Chapter 4), where her fourth graders have been investigating how many different rectangular prisms (boxes) can hold twenty-four chocolates. (For example, a $2 \times 3 \times 4$ box and a $3 \times 8 \times 1$ box both hold twenty-four chocolates. How many more boxes can?) Miki asks them to develop a convincing argument that they have found them all. The children then prepare posters for a gallery walk. (If necessary, students may add to their posters during the walk in order to be better prepared to ask or answer questions during the math congress. Reexamination, rethinking, and revision are all critical ingredients to writing proofs.)

Miki begins the gallery walk by reminding her students about the types of comments that are helpful. "You might write, *I think your strategy is interesting, and I'm trying to figure out why we don't have the same number of boxes,*" she explains. "Or: *It's hard for me to follow your thinking. How do you know you have them all?* Or: *Your poster really convinces me. I agree with your thinking.*" She also suggests that readers notice the way authors begin their proof. "You might comment: *Your strategy is really great. What made you think of starting like that?* Or: *I'm puzzled about this box. Isn't that the same box as this one, just turned around? I need more convincing.* Or: *I understand your first few steps, but I don't know how you got from there to your conclusion.*" She then joins her students on the walk and makes comments and queries on sticky notes as well.

Austin, Gabrielle, and Thomas' poster (see Figure 10.3) exemplifies the beginning of a "proof by cases." They have found all the boxes with one layer and draw them on the lower-right side of their poster. (Note how they use the associative and commutative properties for multiplication, which underlie both their doubling and halving and switching strategies.) During the gallery walk several students place sticky notes on their poster. Gene writes, "I like your work with the layers of one. It's the same as ours. But how do you know that you have them all when you go to more layers?" Gene's question is an honest one. Because his group used layers and was consistent, he is confused by the other group's use of halving and doubling. His question is also powerful, because it gets right to the heart of where they may need to go next. Although they have been systematically examining various halving and doubling strategies, they do not yet have an organizational structure to produce a complete list. However, if they can argue that they have systematically associated and/or commuted the factors, they will have developed a nice proof. Conversely, once Austin's group refines their thinking, their approach may help Gene's group improve their strategy of listing all layers (see Figure 4.3): The fewer number of cases needed in the proof, the more elegant it is!

Danielle and Gabby (see Figure 10.4) have switched the number of layers with the number of rows using the commutative property and associative properties—$(1 \times 24) \times 1 = (1 \times 1) \times 24$—and then listed and drawn six boxes. Michael posts this question on their poster: "We did some flipping like you did but not as much. We did more halving and doubling. Would you get more boxes if you tried halving and doubling?" Michael's question reflects the important role of halving and doubling in his group's work (see Figure 4.2). Children often interpret what other children do in terms of their own work. This question may push Danielle and Gabby to consider these other relationships and whether their collection of six noncongruent boxes is a complete list. The process of halving and doubling is based on factorization, while flipping yields congruent boxes (and illustrates the associative and commutative rules). These are two of the big ideas Miki wants her students to develop.

On her sticky note, Chloe writes, "Right here, I don't exactly get what you did," and places it on the poster where Danielle and Gabby have written, "They are congruent to each other and you still write them differently from each other." This comment prompts Danielle and Gabby to revise their wording about how flipping relates to boxes with different numbers of layers.

Many students, such as Tim, Mary, and Chas (see Figure 4.4), demonstrate they have all the boxes by examining the possibilities within each layer. This is *proof by cases*, and as noted in the discussion of the farm

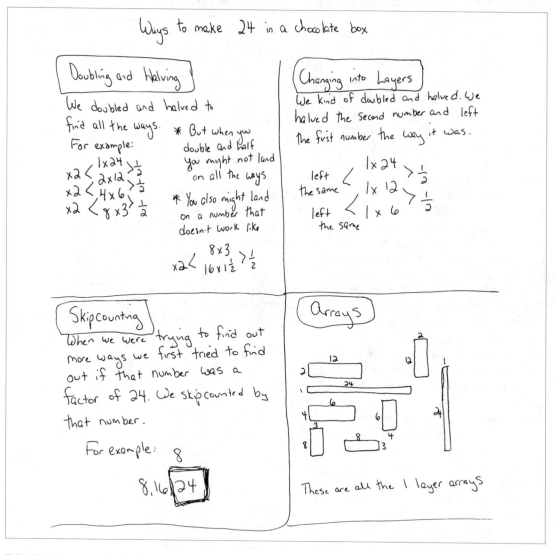

FIGURE 10.3 *Austin, Gabrielle, and Thomas' Poster*

fences work, it typically has two components: showing that you have all the cases (in this instance, determining the possible layers) and then checking that you have what you want in each case. However, other strategies used by these children reveal additional important aspects of the development of proof.

Austin, Gabrielle, and Thomas' early work included many pages in which they listed possible box dimensions using the commutative law and

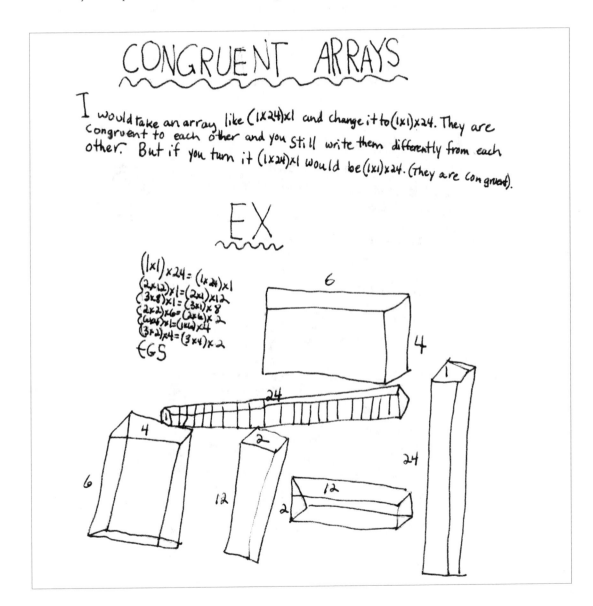

FIGURE 10.4 *Danielle and Gabby's Poster*

doubling and halving strategies. In the top portion of their final poster (see Figure 10.3) they focus on these strategies: "We doubled and halved to find all the ways. But when you double and halve you might not land on all the ways." This is a big idea. They are demonstrating an awareness of a *need for more general argumentation than outlining a process that worked.* Many students were able to double and halve and find what they believed to be a complete list—and for boxes with a volume of twenty-four it is possible to find all possibilities through doubling and halving, and flipping. But

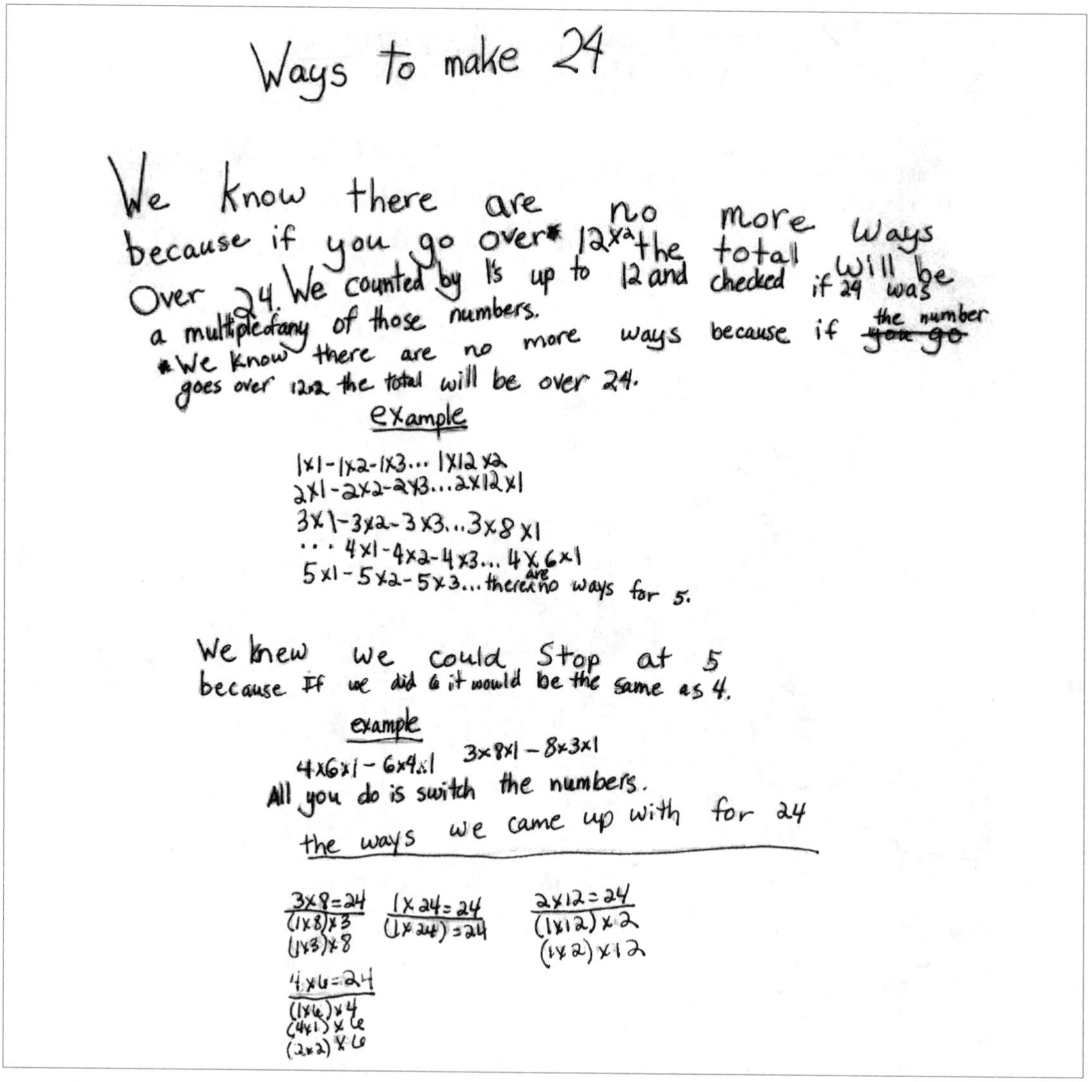

Ways to make 24

We know there are no more ways because if you go over* 12×a the total will be Over 24. We counted by 1s up to 12 and checked if 24 was a multiple of any of those numbers.
*We know there are no more ways because if the number goes over 12×a the total will be over 24.

example

1×1 – 1×2 – 1×3 ... 1×12 ×a
2×1 – 2×2 – 2×3 ... 2×12 ×1
3×1 – 3×2 – 3×3 ... 3×8 ×1
··· 4×1 – 4×2 – 4×3 ... 4×6×1
5×1 – 5×2 – 5×3 ... there are no ways for 5.

We knew we could stop at 5 because if we did 6 it would be the same as 4.

example

4×6×1 – 6×4×1   3×8×1 – 8×3×1
All you do is switch the numbers.

the ways we came up with for 24

3×8=24   1×24=24   2×12=24
(1×8)×3   (1×24)=24   (1×12)×2
(1×3)×8   (1×2)×12

4×6=24
(1×6)×4
(4×1)×6
(2×2)×6

FIGURE 10.5 *Laura and Zach's Poster*

Austin, Gabrielle, and Thomas explicitly raise the need to check that the process will yield all possibilities. A compelling need for careful proof has arisen. They go on to say, "You also might land on a number that doesn't work" and give the example $8 \times 3 \rightarrow 16 \times 1\frac{1}{2}$. Here they are *working for clarity in definitions and processes*—in this context halving and doubling works only when what you need to halve is even.

Laura and Zach's work (see Figure 10.5) is another example of the importance of reexamining one's strategy. They list boxes according to different numbers of layers and then note, "We knew we could stop at 5 because if we did 6 it would be the same as 4. Example: $4 \times 6 \times 1$, $6 \times 4 \times 1$, $3 \times 8 \times 1$, $8 \times 3 \times 1$. All you do is switch the numbers." They have noticed that if they continue to list by layers they will not find any "new" boxes— all they will get are boxes that were listed before (although they would have to be flipped). They have discovered they don't have to make new boxes— they can look at previous entries on the list to get them all. This is a powerful observation that can be an important strategy for condensing a proof. They have *moved beyond a list and are employing general principles to shorten the search*. Although their poster doesn't provide the complete answer, they are examining their thinking carefully and have taken an important step in developing strategies for proving.

WHAT IS REVEALED

It is over time—with opportunities to try to build cohesive arguments— that children come to appreciate the notion of an acceptable argument in a mathematics classroom (Stylianou, Blanton, and Knuth 2009). Unfortunately in most elementary classrooms, children are not given opportunities to consider how to write a mathematical argument for a mathematical audience; they are merely asked to show the teacher what they did and to explain their strategy. They need opportunities to revise their written arguments.

Participating in "gallery walks" implicitly tells children that mathematicians write up their proofs for an audience—an audience that will read these proofs critically. There is no reason to develop a written form of proof without an audience. Often when we have solved a problem and are convinced we are right, it is the process of writing up our proofs that causes us to rethink. And it is the elegancy of the forms we ultimately produce that makes the discipline at once beautiful and creative.

Miki's students are developing proofs by examining cases—most of them classify all the boxes according to the number of layers. This prompts several important developments. Austin, Gabrielle, and Thomas realize that a proof requires more than just retelling what you've done; they begin looking for a more general argument than outlining a process that worked. They also realize they must strive for clarity in what they define. Laura and Zach have found general principles they can use to shorten the list;

although their poster doesn't provide the complete list, these strides are important, as these realizations will make the act of proof more meaningful and accessible.

GENERALIZING FROM SPECIFIC CASES TO ALL CASES

Many mathematical statements that need to be verified involve large collections of numbers. Take "the sum of any two even numbers is an even number." How can this be checked for all pairs of even numbers? Checking an example does not give a proof in all cases. Yet, surprisingly, mathematicians have learned that checking even a single case can lead to a proof *for all* when the conditions are right. A number of researchers have investigated young children's thinking about even and odd numbers (see Schifter 2009, for example). In reasoning the validity of the statement "the sum of two even numbers is an even number," children naturally make a universal generalization from a special case. Typically, they create a diagram of two even numbers, perhaps pictured as two rows of the same length (which they can do because the numbers are even) and then note that when combined you still get two rows of the same length (see Figure 10.6).

The diagram literally shows that 8 + 14 is even. But no matter what even numbers you choose, you can make a version of this same diagram. Therefore, the result is true *for all*. In spite of all our proclamations that one case isn't enough, if the creator of the diagram asserts that the reasoning is general enough to cover all cases, then it is enough. In time these same ideas will be recorded like this: "Since even numbers are multiples of two, if $2x$ and $2y$ are even numbers, then $2x + 2y = 2 (x + y)$ is also an even number." The distributive law, represented as a special case in the diagram, is now being expressed algebraically.

BACK TO THE CLASSROOM

Fifth and sixth graders in a summer intervention program are also using *The Box Factory* (Jensen and Fosnot 2007). Cathy is working alongside the teachers as part of a professional development workshop. The children are exploring the amount of cardboard needed for each of the different boxes, each holding twenty-four oranges. Some students are cutting rectangles from $3/4$-inch graph paper (which matches the size of the multilink cubes

FIGURE 10.6
*Representation That
the Sum of Two Even
Numbers Is Even*

```
X X X X        X X X X X X X
          +                      is even because it is still two identical rows
X X X X        X X X X X X X
```

they used to build the boxes), six rectangles per box to match each side, and then counting the squares. Cathy encourages them to move beyond counting and challenges them to use the array structure and multiplication to determine the area of each rectangular face. Other students just count the squares on each side of each box to determine surface area, and some multiply and then add the six products.

Anthony stares at his list of boxes—$(1 \times 1) \times 24$, $(1 \times 2) \times 12$, $(1 \times 3) \times 8$, $(1 \times 4) \times 6$, $(2 \times 2) \times 6$, $(3 \times 4) \times 2$. He has organized them so he doesn't have to repeat the calculation for the flip of a box—he knows the surface area will be the same if they are flips of each other. He writes:

$$3 \times 2 + 4 \times 2 + 3 \times 2 + 4 \times 2 + 3 \times 4 + 3 \times 4 = 6 + 8 + 6 + 8 + 12 + 12$$
$$= 14 + 14 + 24 = 52$$

His partner, Jeremy, is cutting more graph-paper rectangles. "I'm doing the $(2 \times 6) \times 2$," Jeremy says. "It's the original box."

Anthony looks again at his list and at what Jeremy is cutting out and adds: $2 + 2 \times 6 + 2 + 2 \times 6 + 4 + 4 = 24 + 24 + 8 = 56$. Then, tentatively, he offers a conjecture: "You just go all around where it is the same height and then you do the bottom and top."

"What do you mean?" Jeremy asks with a puzzled look, continuing to cut out six rectangles.

Below his calculation Anthony writes: $l + w \times h + l + w \times h + t + b$. He explains, "We could cut just four rectangles instead of six. There's two rectangles that wrap around the sides. Then we just have to add the top and the bottom."

Jeremy begins to understand the beauty of Anthony's insight. "Oh . . . so you mean we don't need to make all six rectangles. We can put some of them together?"

They continue with this strategy and calculate the amount and cost of the cardboard for each box. As they prepare for the gallery walk, they decide to add the formula they used. In small, almost illegible, print they write $l + w \times 2 \times h + t + b = x$ on the bottom of the poster. During the gallery walk, their poster receives many sticky notes with questions about how they came up with the formula.

In the subsequent math congress, Cathy asks Jeremy and Anthony to discuss the formula. "So what does all this mean that you have here?" She points at the expression with l, w, and h.

"Jeremy was cutting out each side of the box," Anthony begins, "and I noticed that instead of cutting each side out we could think about bigger pieces by putting some of the sides of the box together and we wouldn't have to cut so much. Then I noticed that if we added the length and width twice, that was the length of the piece of paper that went all around. So then just multiply that by the height."

Cathy is not sure all the students are following Anthony's reasoning, particularly his use of the letters l, w, and h, so she asks him to cut out a single paper for the lateral surface area of the $(2 \times 3) \times 4$ box as an example.

"It's going to be two five-by-four rectangles, that's two plus three and then times five for each half, so it's times two." Anthony cuts out the rectangles and holds them up, showing how they make one piece, which wraps around the box. He then describes how a bottom and a top are still needed to make the box. Cathy helps him explain that in this case the l is 2 and the w is 3, then gives the children time to talk with a partner in order to make sense of the discussion themselves. Several classmates ask questions about the height and why he is multiplying by it.

Anthony explains, "The height is the number of layers. Because that is just two sides of the box, I have to double it to account for the other part of the box. Then you add the top and the bottom." On the whiteboard he writes: $l + w \times 2 \times h + t + b = x$.

Marcy, another student, clarifies, "He adds the length and the width, then multiplies it by the height, which is really the number of layers, and adds them up two times."

Cathy asks, "Did he add first or multiply first? How did he do it? Put your thumbs up if you think he added first." Some thumbs go up. "Put your thumbs up if you think he multiplied first." Other thumbs go up.

"I added the length and width first," Anthony clarifies.

Cathy knows that Anthony has invented variables and described his operations symbolically, and that what he wrote makes perfect sense to him, but she is looking for an opportunity to introduce proper order of operations in a natural way. "So, Anthony, since the kids weren't sure, would it be okay if I put parentheses on your formula like this to show you added length and width first?" She does so: $(l + w) \times 2 \times h + t + b = x$. "Mathematicians like to be very clear with what they write, so we put parentheses in like this so people who read the formulas can know what to do first."

Cathy also wants to help Anthony generalize further. She asks, "So you used the length and the width twice. Are the top and the bottom related?"

Anthony says they are the same, and Cathy asks how he would find the area of the top.

"It is l times w," Anthony declares excitedly. He adds parentheses and crosses out the $t + b$. The revised equation now reads:

$$[(l + w) \times 2 \times h] + l \times w \times 2 = x$$

Tim, a visiting teacher, asks Anthony if this equation would give him the surface area of any box, not just a 2, 3, 4.

Anthony thinks for a second and then replies, "Well, what shape is your box? Because if it had triangles on the top and bottom, this wouldn't work. It only works for rectangles."

WHAT IS REVEALED

Based on his experience with a few examples, Anthony has generalized his process, which was to separate finding the lateral surface area from adding the top and bottom. He even created notation to express his ideas (although very likely he had seen $l \times w$ for area before). Aside from the lack of parentheses required for conventional order of operations in his original expression, his formula is correct. More important is his realization that the process is general. Anthony has worked out a specific case and then, realizing there are no constraints on his approach to all other cases, generalizes the result and expresses it symbolically. This is an extremely powerful mathematical observation. Even more is revealed when, answering Tim's question, he articulates his awareness that the box needs to be a rectangular prism—no triangles allowed! This is universal generalization.

We often tell students that "an example is not a proof," because learners tend to overgeneralize in their first attempts at justification. It's important we do so. But we must also be carefully attuned to the nature of their examples and representations—sometimes they will generalize appropriately from a special case. In fact, universal generalization is common throughout mathematics for proving "for all" statements, and Anthony's work, as well as children's work on the sum of even numbers being even, shows that children are capable of formulating such proofs.

SUMMING UP

Proof has taken on different meanings and forms over the history of mathematics. In the early days of the discipline a compelling picture or description was enough for statements to be accepted as "truths." Over the years, though, as mathematicians crafted written forms of their arguments for mathematical audiences, the proof itself became an object of mathematical beauty. Various forms were constructed, such as deductive and inductive proofs, proof by cases, and proof by contradiction. Geometric representations and models became important as well, as did the basic form of reasoning known as *modus ponens*—connecting "if . . . then" statements into a chain of reasoning that holds together without gaps. A second basic rule of inference in mathematical proof is universal generalization—from one case where no special assumptions limit the process to "for all." These forms evolved slowly through the years and were generated through dialogue in a mathematical community.

If we expect children to understand the beauty of proof, we need to create a community of discourse in which they can craft arguments, reflect on these arguments, question them, and revise them. We need to model their thinking and the processes they use, so they have objects to discuss and can examine their logic. Then they can link reasons using deductive rules or talk about why something is always true based on the generality inherent in an example.

Children's "proofs" may look different from the professional mathematician's, but when supported and encouraged to do so, their arguments can take on a surprising logic. Children come to enjoy the process of writing and reading about the mathematical reasoning behind the forms of proof they create, and by doing so they build a foundation for later, more advanced mathematics. They need the opportunity to be seen as young mathematicians hard at work creating their own elegant forms of proof.

REFERENCES

ANGLIN, W. S. 1992. "Mathematics and History." *Mathematical Intelligencer* (14) 4: 6–12.

BLOOM, B. S., J. T. HASTINGS, and G. F. MADAUS. 1971. *Handbook on Formative and Summative Evaluation of Student Learning.* New York: McGraw-Hill.

BOALER, J., and C. HUMPHREYS. 2005. *Connecting Mathematical Ideas: Middle School Video Cases to Support Teaching and Learning.* Westport, CT: Ablex.

CANDLAND, D. K. 1993. *Feral Children and Clever Animals: Reflections on Human Nature.* New York: Oxford University Press.

CARPENTER, T., M. FRANKE, and L. LEVI. 2003. *Thinking Mathematically: Integrating Arithmetic and Algebra in Elementary School.* Portsmouth, NH: Heinemann.

CHANG, M., and C. T. FOSNOT. 2007. *Beads and Shoes, Making Twos.* A Unit of Study in *Contexts for Learning Mathematics.* Portsmouth, NH: firsthand Heinemann.

CHOMSKY, N. 1988. *Language and Problems of Knowledge: The Magna Lectures.* Cambridge, MA: MIT Press.

DEHAENE, S. 1997. *The Number Sense.* New York: Oxford University Press.

DEVLIN, K. 2003. *The Millennium Problems: The Greatest Unsolved Mathematical Puzzles of Our Time.* Cambridge, MA: Perseus.

DEWEY, J. 1902. *The Child and the Curriculum.* Chicago: University of Chicago Press.

DOLK, M., and C. T. FOSNOT. 2006. Multiplication and Division Minilessons: Grades 3–5. Interactive CD. Portsmouth, NH: Heinemann.

DRISCOLL, T. 1999. *Fostering Algebraic Thinking: A Guide for Teachers, Grades 6–10.* Portsmouth, NH: Heinemann.

EVES, H. 1988. *Return to Mathematical Circles.* Boston: Prindle, Weber, and Schmidt.

FOSNOT, C. T. 2007a. *Bunk Beds and Apple Boxes.* A Unit of Study in *Contexts for Learning Mathematics.* Portsmouth, NH: *first*hand Heinemann.

———. 2007b. *Grandma's Necklaces.* A Read Aloud from *Contexts for Learning Mathematics.* Portsmouth, NH: *first*hand Heinemann.

———. 2007c. *The Sleepover.* A Read Aloud from *Contexts for Learning Mathematics.* Portsmouth, NH: *first*hand Heinemann.

FOSNOT, C. T., and A. CAMERON. 2007. *Games for Early Number Sense: A Resource Throughout the Year.* A Unit of Study in *Contexts for Learning Mathematics.* Portsmouth, NH: *first*hand Heinemann.

FOSNOT, C. T., and M. DOLK. 2001a. *Young Mathematicians at Work: Constructing Number Sense, Addition and Subtraction.* Portsmouth, NH: Heinemann.

———. 2001b. *Young Mathematicians at Work: Constructing Multiplication and Division.* Portsmouth, NH: Heinemann.

———. 2002. *Young Mathematicians at Work: Constructing Fractions, Decimals, and Percents.* Portsmouth, NH: Heinemann.

FOSNOT, C. T., and P. LENT. 2007. *Trades, Jumps, and Stops: Early Algebra.* A Unit of Study in *Contexts for Learning Mathematics.* Portsmouth, NH: *first*hand Heinemann.

FOSNOT, C. T. and B. JACOB. 2009. "Young Mathematicians at Work: The Role of Contexts and Models in the Emergence of Proof." In *Teaching and Learning Proof Across the Grades,* edited by D. Stylianou, M. Blanton, and E. Knuth. New York: Routledge.

FREUDENTHAL, H. 1973. *Mathematics as an Educational Task.* Dordrecht: Reidel.

———. 1991. *Revisiting Mathematics Education: The China Lectures.* Dordrecht: Kluwer.

GAGNE, R. 1965. *The Conditions of Learning.* London: Holt, Rinehart, and Winston.

GOLDENBURG, E. P., and SHTEINGOLD, N. 2008. "Early Algebra: The Math Workshop Perspective." Chapter Seventeen in *Algebra in the Early Grades,* edited by J. J. Kaput, D. W. Carraher, and M. L. Blanton. Mahwah, NJ: Lawrence Erlbaum Associates.

GRAVEMEIJER, K. 1994. *Developing Realistic Mathematics Education.* Utrecht, The Netherlands: Freudenthal Institute.

———. 1999. "How Emergent Models May Foster the Constitution of Formal Mathematics." *Mathematical Thinking and Learning,* I (2): 155–77.

GRAVEMEIJER, K., and F. VAN GALEN. 2003. "Facts and Algorithms as Products of Students' Own Mathematical Activity." In *A Research Companion to Principles and Standards for School Mathematics,* edited by J. Kilpatrick, W. G. Martin, and D. Schifter, 114–22. Reston: NCTM.

GUEDJ, D. 1997. *Numbers: The Universal Language.* New York: Harry Abrams.

Izard, V., P. Pica, E. Spelke, and S. Dehaene. 2008. "The Mapping of Numbers on Space: Evidence for an Original Logarithmic Intuition." *Medical Science* (Paris). Dec; 24 (12):1014–6.

Jacob, B., and C. T. Fosnot. 2007. *The California Frog Jumping Contest*. A Unit of Study within *Contexts for Learning Mathematics*. Portsmouth, NH: *firsthand* Heinemann.

Jensen, M., and C. T. Fosnot. 2007. *The Box Factory*. A Unit of Study within *Contexts for Learning Mathematics*. Portsmouth, NH: *firsthand* Heinemann.

Kamii, C. 1985. *Young Children Reinvent Arithmetic*. New York: Teachers College Press.

Kaput, J. J., D. W. Carraher, and M. L. Blanton, eds. 2008. *Algebra in the Early Grades*. Mahwah, NJ: Lawrence Erlbaum Associates.

Lager, C. 2007. "Reading Interactions That Unnecessarily Hinder Algebra Learning and Assessment." Unpublished document, University of California, Santa Barbara.

Lent, P., E. Wall, and C. T. Fosnot. 2006. "Young Mathematicians at Work: Constructing Algebra in Grade Two." *Connect*. (19) 3, January/February.

Meyer, M. 2001. "Representations in Realistic Mathematics Education." In *The Roles of Representation in School Mathematics, 2001 NCTM Yearbook*. Reston, VA: NCTM.

National Research Council. 2001. *Adding It Up: Helping Children Learn Mathematics*. Washington, DC: National Academy Press.

NCTM. 1989, 2000. *Principles and Standards for School Mathematics*. Reston, VA.

Piaget, J. 1977. *The Development of Thought: Equilibration of Cognitive Structures*. New York: Viking.

Schifter, D. 2009. "Representation-based Proof in the Elementary Grades." In *Teaching and Learning Proof Across the Grades*, edited by D. Stylianou, M. Blanton, and E. Knuth. New York: Routledge.

Schifter, D., and C. T. Fosnot. 1993. *Reconstructing Mathematics Education: Stories of Teachers and Children Meeting the Challenge of Reform*. New York: Teachers College Press.

Schifter, D., S. J. Russell, and V. Bastable. 2006. "Is it 2 More or 2 Less? Algebra in the Elementary Classroom." *Connect*. (19) 3, January/February.

Stewart, I. 2008. *The Story of Mathematics: From Babylonian Numerals to Chaos Theory*. London: Quercas.

Stylianou, D., M. Blanton, and E. Knuth. 2009. *Teaching and Learning Proof Across the Grades* (Studies in Mathematical Thinking and Learning Series). New York: Routledge.

Treffers, A. 1987. *Three Dimensions: A Model of Goal and Theory Description in Mathematics Instruction—the Wiskobas Project*. Dordrecht: Reidel.

VAN AMERON, B. 2002. *Reinvention of Early Algebra.* University of Utrecht: Freudenthal Insitute.

VAN REEUWIJK, M. 1995. "The Role of Realistic Situations in Developing Tools for Solving Systems of Equations." Paper presented at the Symposium on the Learning and Teaching of Algebra. AERA, San Francisco.

WYNN, K. 1998. "Psychological Foundations of Number: Numerical Competence in Human Infants." *Trends in Cognitive Sciences*, (2): 296–303.

INDEX

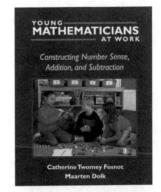

The *Contexts for Learning Mathematics* series by **Catherine Fosnot** and her colleagues from Mathematics in the City and the Freudenthal Institute uses carefully crafted math situations to foster a deep conceptual understanding of essential mathematical ideas, strategies, and models. The series' 18 classroom-tested units are organized into 3 age-appropriate packages.

Number Sense, Addition, and Subtraction Grades K–3
978-0-325-01052-6 / 2007 / 12 Books + 8 Read-Aloud Books + CD-ROM / **$206.00**

Multiplication and Division Grades 3–5
978-0-325-01053-3 / 2007 / 8 Books + 17 Posters + CD-ROM / **$166.00**

Fractions, Decimals, and Percents Grades 4–6
978-0-325-01054-0 / 2007 / 7 Books + 16 Posters + CD-ROM / **$166.00**

Fosnot's Complete Elementary Bundle
978-0-325-01004-5 / 2007 / K–3 + 3–5 + 4–6 / a $538.00 value for only **$484.20**

Complete content-based packages are ideal for math coaches, lead teachers, and resource libraries. Teacher packs and individual titles are also available for purchase.

For more information visit **contextsforlearning.com**